Challenges and Opportunities
in
Nanotechonlogy

Challenges and Opportunities in Nanotechonlogy

Editor

B.K. Parthasarathy

ISHA

ISHA BOOKS

Delhi-110009

Challenges and Opportunities in Nanotechnology

ISBN: 81-8205-433-8

Published in 2007 in India by
ISHA Books
B-63, 1st Floor, New Gupta Colony,
Near DDA Market,
Delhi-110009
Phone: 9212142040
ishabooks@hotmail.com

Laser Type Setting by: Rudra Computer Graphics, Delhi
Printed at: Salasar Imaging System, Delhi

Preface

No other science which has emerged within the latter half of the 20th century has perhaps garnered so much attention and fascination as nanotechnology. Embodying the adage"small is beautiful" within its essence and principles, nanotechnology is involved in the control of matter in the range of nanometres, and the fabrication of devices along the same scale. One can almost say that it seeks to miniaturise the devices and technology which form a part of our scientific existence, in fact, miniaturise life itself. It holds immense potential to revolutionise the way we lead our lives, and is already being heralded as a New Age Renaissance-like scientific development.

Nanotechnology is the science which has already made great inroads in the field of scientific human endeavour, and shows no signs of relenting this position. A highly multidisciplinary field, it draws from such fields as electrical engineering, colloidal chemistry, device physics, etc. Originating as a manifestation of a talk given by a visionary scientist at an after-dinner talk, nanotechnology has come a long way in proving itself as one of the most formidable scientific branches to ever emerge, holding the power to completely alter the course of human civilisation. It is already being used widely in areas as diverse as pharmaceuticals, engineering, medicine, cosmetics, weapons, etc. Nonetheless, inspite of the tremendous benefits which are staring at humanity through this science, sceptics and their reservations abound, and rightly so.

The implications of nanotechnology are multifold, not the least among them being that in human health and the environment. Many cite nanoparticles as being toxic, and the smaller these particles get, the bigger the threat they pose in this regard, for the large surface area to volume ratio implies that the biological and chemical reactivity of nanoparticles would be greater, making them highly dangerous to humans and their environment. These, however, are only short-term worries; the long-term worries are far more greater and sociological in nature. Social scientists predict that the advent of this form of technology can very well serve to aid the already wide gap between the world's richest and the poorest, forming a kind of nano-divide, and may even lead to a post-scarcity economy. All these factors balance the scales highly in favour of the sceptics, and nano-enthusiasts may find it hard to support their cause in the light of such strong arguments.

This book has been written with the clear purpose of presenting forth a forum wherein the discussion goes behind the hype generated by nanotechnology, and explores the darker side of this science. The above mentioned factors make it highly crucial that nanotechnology prepare itself to not just revolutionise technology, but also brace itself against the challenges presented by them, to seek opportunities to work upon them, and keep them at bay. Through incisive, critical studies, the book examines in an in-depth manner the problems faced by nanotechnology, and how it may equip themselves against them. The emphasis on what reports and studies which have predicted such grim outcomes of a full-scale advent of this science, and what they foretell for the future. The tone of the book is highly inquisitive and searching, and it is hoped that this serves well for the readers interested in knowing more about nanotechnology.

Editor

Contents

Preface *v*

1. Introduction 1

2. Nanoscale Architecture 19

3. Developments in Nanotechnology 65

4. Nanotechnology and Industrial Development 91

5. Nanotechnology Applications in the Energy Sector 118

6. Nanoelectronics: Opportunities and Challenges 166

7. Economic Impact of Nanotechnology 202

8. Public Concerns about Nanotechnology 226

9. Future of Nanotech Revolution 252

Bibliography 295

Index 297

1 Introduction

The term 'nanotechnologies' encompasses an even wider range of basic science, methods and engineering approaches than biotechnology, and so are likely to give rise to a highly diverse set of potential applications over very different time-scales.

Nanomaterials, which are materials that have one or more dimensions in the range of 1 to 100 nanometers. The reason that size matters is that the properties of materials can have some unexpected differences from their behaviour in larger bulk forms that makes for new application opportunities.

The two reasons for this change in behaviour are an increased relative surface area (producing increased chemical reactivity) and the increasing dominance of quantum effects. In conventional technology, materials are a critical if sometimes underappreciated component in achieving the goals of nanotechnology.

A central construct of materials science and engineering is that the structure and processing of a material give rise to its properties, which in turn determine its performance in use. This certainly applies to nanomaterials as well and provides a useful framework for understanding in a familiar context.

Nanomaterials have been important in the materials field for quite a long time. An early example was the incorporation of gold nanoparticles in stained glass in 10 A.D. At sizes in the nanorange, gold can exhibit a range of colours. Also, nano-sized carbon black particles have been used to reinforce tires for 100 years.

Perhaps one manifestation of the concept of nanomaterials familiar to traditional metallurgists is precipitation hardening. As discussed by Hornbogen, the accidental discovery of precipitation hardening in 1906 by Wilm in Duralumin alloys opened the way to significant improvements in strength properties for aluminum.

Although researchers were unable to image the precipitates with the instruments available at the time, it was subsequently discovered that nanometersised precipitates were the source of hardening. Through the advent of electron microscopy came a better understanding of the structure of precipitates, including aspects such as size and coherency with the metal matrix.

As a framework for this discussion, the approach of R. Jones for organising nanotechnology into three categories can be applied to nanomaterials. The categories are:

— Incremental nanotechnology-improving the properties of materials by controlling their nanoscale structure.

— Evolutionary nanotechnology-taking a step beyond redesigning simple materials at the nanoscale and designing nanoscale devices that do something interesting

— Radical nanotechnology-developing nanoscale machines that would exist at the convergence of nanotechnology, biotechnology, information technology, and cognitive technology.

Incremental Nanotechnology

Incremental nanomaterials are materials that have improved properties at the nanoscale, typically as a result of their greatly increased surface area, but do not typically take advantage of the quantum effects. Nanoparticles are already seeing application, taking advantage primarily of the high surface area of these fine powders. Nanoceramic powders, the most commercially important of which are simple metal oxides, constitute almost 90% of the total market. For example, nano-sized zinc oxide particles are in use in sunscreen. Nanostructured ceramic coatings are adding durability and toughness to hulls of U.S. Navy ships.

Metal powders are important, as well. Iron nanoparticles have been used to treat groundwater contaminated with trichloroethylene while aluminum nanoparticles have been developed that, due to their increased surface area, have substantially greater "bang for the buck" as solid rocket propellant. Incremental nanomaterials also include polymer nanocomposites in which clay nanoparticles are incorporated to increase the hardness and reduce the permeability of the polymer.

These have seen application in automotive panels and step assists in vans. Other examples of these types of applications are nanoparticles in tennis balls and carbon nanotubes in tennis racquets. Incremental nanotechnology can also be found in the development of nanostructured materials by techniques such as equal channel angular extrusion and accumulative roll bonding, which fall in the category of severe plastic deformation processes.

The primary benefit of these processes is to produce ultrafine-grained structures with dimensions in the submicrometer range, resulting in increased strengthening along with other enhanced properties. Note again that the benefits obtained in these nanostructured materials come primarily from the increase in surface area and the impact of the small scale of the structure on dislocation motion rather than quantum effects. Also note that the inherent coupling of the ability to image these structures and their behaviour is inextricably linked to their optimisation.

Evolutionary Nanotechnology

Evolutionary nanotechnology takes advantage of the changes that can occur in materials at the nanoscale related both to increased chemical reactivity and the increasing importance of quantum effects. Examples include nanoscale sensors that exploit the large surface area of nanotubes and semiconductor nanostructures such as quantum dots and quantum wells.

Silica-based nanomaterials, molecular imprinted polymers, and silicon platforms are envisioned for collection, concentration, and detection of chemical weapons and other related compounds in security and defence applications. The biggest steps are being made in evolutionary nanotechnology, with more and more products expected to appear in the market in the next five years. A material that has generated substantial interest is the class of molecularly perfect carbon structures called fullerenes.

The creation of a 60 carbon atom molecule in the form of a geodesic sphere, termed the buckyball by Smalley in 1985, has lead to other fullerenes containing many more atoms of carbon and taking different shapes. One such shape is the carbon nanotube (CNT), which has a range of unique properties. These materials can act as either highly conductive nanowires or as semiconductors, depending on the specific arrangement of the carbon atoms.

The perfection of the CNT creates a material with a strength reported to be 30 times that of steel at 1/6 the density. The thermal conductivity of CNTs is 50% higher than that of diamond. These unique properties have already led to the use of CNTs in specialty applications such as high-performance tennis racquets and selected electronics. Longer term, CNTs are seen as key components in fuel cells, electrical transmission lines, and thermoelectric conversion devices.

Carbon nanotubes are by no means the sole materials focus of evolutionary nanotechnology. Other materials are in development that will generate immediate interest and impact on the electronic materials field. Medical applications enabled by nanomaterials including diagnosis and treatment, along with markedly different approaches to pharmaceuticals, are also predicted to be on the horizon in the next decade.

Radical nanotechnology is viewed as having an impact much farther into the future. Examples include nextgeneration military uniforms being developed by the Massachusetts Institute of Technology's Institute for Soldiering Nanotechnology.

The uniforms are designed to defend against chemical and biological weapons, provide ballistic protection, monitor health, administer medical aid, and provide communication capabilities. Much farther in the future are concepts in which nanotechnology converges with biology, information technology, and cognition.

Since nanomaterials have little mass and are dominated by surface area and size effects, the processes and equipment for nanotechnology-based manufacturing are expected to differ significantly from those currently used. One of the key differences is the concept of manufacturing on an atom-by-atom basis, which has been labelled bottom-up processing.

Conventional methods, called top-down techniques, start with a block of material, etching or milling it down to the desired shape. The main challenge for top-down manufacture is the creation of increasingly small structures with sufficient accuracy, whereas for bottom-up manufacture, the challenge is to make structures large enough, and of sufficient quality, for use as materials.

These two methods have evolved separately and have now reached the point where the best achievable feature size for each technique is approximately the same. Most of the current nanomaterials with potential for structural use are powder based, produced from processes such as inert-gas condensation, electrodeposition, atomisation, and mechanical alloying. They need to be consolidated into shapes with minimal porosity without creating significant structural coarsening and loss of the desirable nanostructural properties. This requires a balance of minimal high-temperature exposure and adequate pressure, and innovative methods of consolidation need to be developed to produce successful materials. While nanotubes have incredible strength, capturing this strength in a structural component is no small challenge.

This challenge is analogous to the challenges faced by early metalmatrix composite materials engineers who sought to capture the ultra-high strength of ceramic whiskers in structural materials. Incorporating these expensive reinforcement materials cost-effectively in large-volume operations was difficult, often resulting in damage to the whiskers and loss of much of their property advantage.

Billions of dollars are being invested in opportunities centered around materials that exist on the scale of a billionth of a meter. In US, the National Science Foundation has made the bold prediction that the business impact of nanotechnology will be $1 trillion by 2015. Some 800,000 to 2 million new jobs, in a field where the estimates of current number of workers range from 20,000 to 40,000, will be required to reach this level.

Pundits claim that within the next decade, nanotech will have huge effects on many practical industries, including manufacturing, health care, energy, agriculture, communication, transportation, and electronics. Just issued, The Nanotech Report 2004 produced by Lux Research highlights the magnitude of financial resources being put into nanotechnology. Specifically, the report notes:

— Governments, corporations, and venture capitalists will spend more than $8.6 billion worldwide on nanotechnology R&D in 2004.

— National and local governments across the world will invest more than $4.6 billion in nanotechnology R&D in 2004. Roughly 35% of this investment is in North America, 35% in Asia, 28% in Europe, and 2% in the rest of the world.

— The U.S. government will spend nearly twice as much on nanotechnology this year as it did on the Human Genome Project (HGP) in its peak year. In 2005, the National Nanotechnology Initiative will surpass the HGP on a cumulative basis. The United States has appropriated more than $3.16 billion for nanotechnology R&D since 2000 and is proposing $982 million in new funding for 2005.

— Established corporations will spend more than $3.8 billion globally on nanotechnology R&D in 2004.

It is not an exaggeration to say that nanotechnology is the biggest thing to happen to the physical sciences for quite some time. It is estimated that worldwide government funding in nanotechnology increased by seven times from 1997 to 2003, exceeding $3 billion in 2003.

In the United States, the 21st Century Nanotechnology Research and Development Act, signed in December 2003, will provide $3.7 billion for four years beginning in 2005. The funding request for $982 million in 2005 is an increase of more than 500% compared to 2001. A key component of this funding is for five user facilities, called the Nanoscale Science Research Centers, which are now under construction at Department of Energy laboratory sites.

These research facilities will focus on synthesis, processing, and fabrication of nanoscale materials. They will be co-located with existing user facilities to provide sophisticated characterisation and analysis capabilities. Also included is funding for addressing so-called grand challenge areas.

Three of these areas, Nanostructured Materials by Design, Manufacturing at the Nanoscale, and Nano-Electronics, -Photonics, and Magnetics, have a materials focus.

Another indicator of the potential in nanotechnology is indicated by the growth in patent activity. In the United States, for example, the number of patents issued in areas such as atomic force microscopy and quantum dots increased ten-fold from 1994-2003.

The U.S. Patent and Trademark Office has been prompted to create a classification system for nanotechnology so that it can track the number, pendency, and assignment of nanotechnology patent applications. While at present there is a classification for fullerenerelated patent applications there are no separate classes for areas such as nanotubes or nanowires.

Also harkening the transition from laboratory to commercialisation is the presence of organisations such as the NanoBusiness Alliance, which claims to be the first industry association founded to advance the emerging business of nanotechnology and microsystems. The NanoBusiness Alliance's mission is to create a collective voice for the emerging small-tech industry and develop a range of initiatives to support and strengthen the nanotechnology business community.

The business prospects of nanotechnology have caught the attention of Wall Street, where brokerage firm Merrill Lynch has been promoting a nanotech stock index. A nanotechnology company, Nanosys, contemplated an initial public offering of stock but withdrew it, which was viewed in a Wall Street Journal article as "less an indictment of nanotechnology than a sign of how tough it is for any company in an early stage of development right now."

The article goes on to note, however, that "nanotechnology remains difficult to define, causing additional uncertainties for investors. . . Some companies have used the nanotechnology label to hype unrelated products, while many real advances are occurring inside big companies such as Intel, where the developments have only a modest impact on stock prices."

Nanomaterials are one area where real revenues are being realised currently. It is estimated that sales of nanomaterials were $1.5 million in 1999 and grew to $430 million by the end of 2003. With this annual growth rate of 300%, the market size would be projected to be $1.3 billion in 2004 and $5 billion in 2005. A more conservative estimate of $1 billion in revenue in 2007 is provided buy a recent study by the Freedonia Group.

Nanotechnology is already having an impact on the structure of the scientific community. The intersection of physics, chemistry, and materials science at the nanoscale has required a broadening of the educational process, which will be required to meet the need for future nanotechnologists. More than 300 academic programs in nanotechnology exist today, with at least 200 in the United States and 100 internationally.

There are implications for professional societies as well, and it has been noted that the scientific and engineering communities should create new means of interdisciplinary training and communication, reduce the barriers that inhibit individuals from working across disciplines, develop links to a variety of other technical and medical organisations, and address ethical issues related to technological developments.

Nanotechnology is gaining awareness beyond the scientific community. The number of times terms related to "nano" appeared in the popular press has risen from 190 in 1995 to a projection of more than 12,000 in 2004.

Recent studies carried out both in the United States and United Kingdom indicated that the public is not as aware yet of nanotechnology. Of those polled in the United States, 71% had heard "little" or "nothing" about nanotechnology while only 29% in the United Kingdom had heard of it. Of that 29%, only 19% were able to offer some sort of definition.

Issues in Nanotechnology

Some currently envisioned applications of nanotechnologies which are seen as technically feasible may never be realised on a significant scale, while unanticipated scientific breakthroughs may lead to rapid applications that are not currently foreseen.

Most of the social and ethical issues arising from applications of nanotechnologies will not be new or unique to nanotechnologies. Nanotechnologies, whether singly or in convergence with other technologies, are likely to hold a range of both positive and (however unintended) negative outcomes. Widespread acceptance and use of nanotechnologies will depend upon a range of social factors including: specific technical and investment factors; consumer choice and wider public acceptability; the political and macro-economic decisions that contribute to the development of major technologies and outcomes that are viewed as desirable; and legal and regulatory frameworks.

Equally, just as the knowledge-base underpinning science and technology can change rapidly and unpredictably, so can society. Forecasting people's needs and values 20 years or more into the future is fraught with uncertainty. Accordingly, it is difficult to state with any confidence how nanotechnologies are likely develop in the future, in interaction with a changing society and its shifting social and ethical concerns. It may also be important to look beyond the perspective of Western industrialised societies, to take account of the ways in which people in developing societies might respond to developments in nanotechnologies and their impacts. Precisely which social and ethical issues become a focus of concern will hinge upon the actual trajectories of change in particular nanotechnologies.

Wood et al point out that current evaluations of the impacts of nanotechnologies can be located on a continuum whose extreme poles are on the one hand incremental progress, and on the other a radical disjunction from current science and technology. Equally, much of what passes for incremental nanotechnologies may still prove transformative in social terms. This is because the role of nanotechnologies is likely to be an enabling one, often in convergence with other new technologies, bringing to fruition applications such as pervasive sensing which have been anticipated by commentators for many years, but are only now becoming a practical possibility.

Much of the commentary provided to the working group also suggests that many of the social and ethical issues raised by incremental developments are unlikely to prove entirely new. For example, many concerns about the overall impact of a rapidly changing science on society, and the governance and regulation of the technology, are likely to echo some of those raised previously about other developments in science and technology that have proved controversial, such as nuclear energy, reproductive technologies or biotechnology. That does not make these concerns any the less significant or worthy of the attention of policy-makers; indeed, past experience with controversial technologies should predispose policymakers to pay timely and applied attention to these concerns rather than dismissing them as 'nothing new'.

There is some disagreement about how much of an economic impact nanotechnologies will have. A range of views has been heard in evidence. Overall, it seems that effects will be incremental in the short term but, given the fact that nanotechnologies are likely to enable a great many products and processes, they may well have significant economic impact in the long-term.

As argued above, much will depend upon which particular applications eventually come to market, and the order in which they are developed. Judging by experience, seemingly insignificant technological advances could have profound long-term economic impacts. Some commentators appear to assume that the potential economic results of nanotechnologies—greater gross domestic product (GDP), greater efficiency and less wastage in industrial processing—will be entirely positive across society or across the range of developed and developing nations. However, in general the introduction of new technologies creates both 'winners' and 'losers'; for example, as employment is displaced from one sector to another. At this stage, evidence does not suggest that nanotechnologies raise economic issues that differ significantly from other cases of technological innovation.

However, it would contribute greatly to the wider societal debate and to decisions about the introduction of nanotechnologies if appropriate economic analysis of developments with widespread societal impacts, including an assessment of the advantages and disadvantages, is undertaken at an appropriate stage. Since this cannot be done on a systematic basis far ahead of the development of new technologies, such analysis would need to proceed on a caseby-case basis, as developments and applications come closer to market. Such analysis should also take full account of the uncertainties involved, of the case for relying on alternative technologies and of any economic shocks and surprises. The application of science, technology and engineering has undoubtedly improved life expectancy and quality of life for many in the long term. In the short-term, however, technological developments have not necessarily benefited all of humankind, and some have generated very definite 'winners' and 'losers'.

Concerns have been raised over the potential for nanotechnologies to intensify the gap between rich and poor countries because of their different capacities to develop and exploit nanotechnologies, leading to a so-called 'nanodivide'. If global economic progress in producing high-value products and services depends upon exploiting scientific knowledge, the high entry price for new procedures and skills (for example, in the medical domain) is very likely to exacerbate existing divisions between rich and poor.

Equally, a parallel danger that could arise if the more radical 'visions' of the promise of nanotechnologies were realised, is that enthusiasm for developing a 'technical fix' to a range of global and societal ills might obscure or divert investment from cheaper, more sustainable, or low-technology solutions to health and environmental problems. A further concern that has been highlighted relates to patenting in the field of nanotechnologies.

Appropriate ownership of intellectual property is generally considered advantageous. However, as experience in genetics shows, patents that are too broad or that do not strictly meet the criteria of novelty and non-obviousness, can work against the public good. There is a concern that broad patents could be granted on nanotechnologies, for example on processes for manipulating or creating materials, which would stifle innovation and hinder access to information, not least by those in the developing world.

There have also been suggestions that nanotechnologies could offer new opportunities for some developing countries to participate more directly in global technology through their own initiatives—for example, through the development of plastic electronics facilities, which are one-hundredth of the cost of conventional silicon fabrication plants.

There are therefore significant risks that some short-term developments in nanotechnologies will be exclusive to those who already own wealth and power, to the detriment of wider society. Equally, opportunity to apply nanotechnologies in ways that will benefit the developing world should not be overlooked. The governance of nanotechnologies must in some way be designed to incorporate the perspectives and objectives of governments, the market and civil society.

Enhancement Applications

Nanotechnologies are contributing to the development of some 'enhancement' applications; the closest to development being improved cochlear and retinal implants, to improve or restore hearing and eyesight. A few disability rights groups have objected to proposed interventions that enhance human capacities, on the grounds that this might lead to stigmatisation of those without enhanced capacities. The general concern is difficult to grasp without a clear account of the difference between enhancements and other interventions.

The issue of specific human enhancements is also likely to fall, initially at least, squarely within the medical domain, where there is an established history of considering emergent ethical issues and the societal acceptability of particular procedures. The general issues about stigmatisation of those who are different in various ways is a serious one, but it has little connection with ways in which differences between people may be brought about.

All successful medical treatment of illness, including treatment of illness with a genetic basis, enhances the functioning and capacities of those who are treated. Even where an intervention—a drug, a prosthesis, a medical device, surgery—is not effective for all sufferers, it can hardly be withheld from those who could benefit on the grounds that some others cannot. There is no general case for resisting technologies or interventions that enhance human capacities. It would be wrong to deny appropriate treatment to patients whose impaired sight can be improved by glasses or surgery simply because others have sight impairments that cannot be improved. However, certain types of enhancement may be more controversial, whether because those who lack them would be stigmatised or (more probably) for other reasons.

Some have argued that all enhancement by gene therapy is an unacceptable form of eugenics, while others have argued that genetic enhancement of basic capacities such as intelligence or height would only be acceptable only if fairly distributed. Yet others hold that if enhancement of capacities by education or training is acceptable, then enhancement of capacities by other means, such as cosmetic surgery or taking drugs with cognitive effects, is also acceptable.

Convergence of Nanotechnologies

Convergence refers to the multiple ways in which nanotechnologies will combine in the future with other developments in new technology. Convergence probably presents some of the biggest uncertainties, with respect to what is genuinely plausible and when new technologies might actually come into use.

Convergence of nanotechnologies with information technologies could raise concerns about civil liberties. However, convergence is likely to generate a range of other social and ethical challenges, particularly in relation to longerterm applications within bio-nanotechnology that involve significant interface of material systems with, or internal modification of, the body. Some developments—although essentially physical interventions conducted primarily for medical benefits - might well raise a range of fundamental psychological and sociological questions centred around the issue of identity: that is, what we understand to be 'human', what is 'normal' and what is not.

As stated in the recent report from the German Parliament Office of Technology Assessment: 'In visions of nanotechnology, we repeatedly see aspects which dissolve the boundaries between what constitutes a human being, and what they can create with the help of technological achievements and applications. Such aspects relate for example to the penetration and modification of the human body by attempts to supplement or replace its biological components by nanotechnology components, and to network it with external machnes or other bodies or body parts'. Developments that in some way invade or intervene with the body in the manner described above are also likely to raise issues of control and choice and to be particularly sensitive in relation to public perceptions and concern. Stephen Wood and Richard Jones highlighted that although these very extreme visions of the potential outcomes of nanotechnologies - including the possibility of greatly expanding lifespans, or even of the separation of human consciousness from the body and its relocation in a computer—may seem too far-fetched for many scientists, these visions do form a background for discussions of the impact of nanotechnologies by informed non-scientists.

Defence Capability

Nanotechnologies are predicted to offer significant advances and advantages in defence capability. According to the UK Ministry of Defence, nanotechnologies will present both new opportunities for defence and new external threats. Echoing the points made above about the prospects for the development of pervasive sensing, the main initial defence impact is predicted to be in information systems using large numbers new and cheap sensors, as well as in information processing and communications.

These developments might enable pervasive nanosensors to contribute to national defence capability through early detection of chemical or biological releases, and increased surveillance capability. In addition, 'a whole range of military equipment including clothing, armour, weapons, personal communications will, thanks to low cost but powerful sensing and processing, be able to optimise their characteristics, operation and performance to meet changing conditions automatically'.A current military example is provided by the US Institute for Soldier Nanotechnologies at Massachusetts Institute of Technology, which has been awarded a $50 million budget from the US army to research new materials.

Its ultimate goal is 'to create a 21st century battlesuit that combines high-tech capabilities with light weight and comfort', focusing on soldier protection, injury intervention and cure, and human performance improvement. Military developments raise several obvious social and ethical issues, most of them once again not confined to nanotechnologies. Manipulation of biological and chemical agents using nanotechnologies could result in entirely new threats that might be hard to detect and counter. Some observers have suggested that refinements of both existing and new weapons systems, through applications of nanotechnologies, might lead to a new form of arms race.

One can also ask whether the use of arms control frameworks developed for existing categories of nuclear, chemical and biological weapon will be sufficient to control future developments involving nanotechnologies. A related issue arises from the fact that much of the basic knowledge and technology needed to achieve military capabilities using applications of nanotechnologies will be produced within the civil sector, and hence is potentially available to a very wide range of parties, including non-state actors. Joy suggested that 'The 21st-century technologies -genetics, nanotechnology, and robotics (GNR)—are so powerful that they can spawn whole new classes of accidents and abuses.

Most dangerously, for the first time, these accidents and abuses are widely within the reach of individuals or small groups. They will not require large facilities or rare raw materials. Knowledge alone will enable the use of them'. This factor also makes proliferation of weapons development programmes much harder to detect because the line between non-military and military industrial activity becomes blurred.

In this way, nanotechnologies may increase the range of asymmetric power relations. Those applications of nanotechnologies that attract military funding are likely to raise other concerns: for example considerations of secrecy will make the open peer review of findings in these areas much more difficult.

An unintended consequence of secrecy in the development of some nanotechnologies could also be to fuel public distrust and concerns about non-military developments. This would be so particularly if the term 'nanotechnology' as a whole became to be closely associated with military ends. The case of nuclear energy is instructive here. Flynn in the USA argues that one of the historical reasons for the stigmatisation of, and enduring hostile public attitudes towards, nuclear power was the inability of the civilian nuclear industry to separate itself from destructive uses of the atom.

Government denials of this linkage—scarcely believed at the time—further served to undermine public trust in those regulating the technology. There seems to be a significant danger that public acceptance of a whole range of beneficial applications of nanotechnologies, particularly in the environmental domain, might be threatened by too close an association with military applications.

2

Nanoscale Architecture

Nanotechnology is the design, fabrication and use of nanostructured systems, and the growing, shaping or assembling of such systems either mechanically, chemically or biologically to form nanoscale architectures, systems and devices. The original vision of Richard Feynman was of the 'bottom-up' approach of fabricating materials and devices at the atomic or molecular scale, possibly using methods of self-organisation and selfassembly of the individual building blocks. An alternative 'top-down' approach is the ultraminiaturisation or etching/milling of smaller structures from larger ones. Both approaches require a means of visualising, measuring andmanipulating the properties of nanostructures; computer-based simulations of the behaviour of materials at these length scales are also necessary.

Atoms and Solids

The structure of the atom arises as a direct result of the wave-particle duality of electrons, which is summarised in the de Broglie relationship, $\lambda = h/m_e v$, where λ is the (electron) wavelength, m_e is the (electron) mass, v is the velocity and $h = 6.63 \times 10^{-34}$ Js is the Planck constant. The wave-particle duality of the electron means that an electron behaves both as a wave and a particle. This is conveniently summarised in the idea of a wave packet a localised wave that is effectively the summation of a number of different waves of slightly differing wavelengths.

Using these ideas we come to our first model of the atom, the Rutherford-Bohr model. Here the small central nucleus of the atom consists of positively charged protons and (neutral) neutrons. Electrons orbit the nucleus in stable orbits. The allowed, stable orbits are those in which the electron wavelength, given by the de Broglie formula, is an integral multiple *n* of the circumference of the orbit *r*:

$$2\pi r = n\lambda = \frac{nh}{m_e v}. \tag{1}$$

This implies that

$$m_e vr = \frac{nh}{2\pi}, \tag{2}$$

in otherwords, the angular momentum mevr is quantised in that it is an integral multiple of $h/2\pi$. The Bohr model leads to the idea that only certain electron orbits or shells are allowed by this quantisation of angular momentum (i.e., the value of *n*). The Bohr shells in an atom are labelled according to the quantum number, *n*, and are given the spectroscopic labels K, L, M, N, etc. To understand the form of the periodic table of elements, it is necessary to assume that each Bohr shell can contain $2n^2$ electrons.

For instance, a K shell (n = 1) can contain 2 electrons, whereas an L shell ($n = 2$) can accommodate 8 electrons. As well as having a distinct form and occupancy, each shell also has a corresponding well-defined energy. It is usual to define the zero of the energy scale (known as the vacuum level) as the potential energy of a free electron far from the atom. In order to correspond with atomic emission spectra measured experimentally, the energies of these levels E_n are then negative (i.e., the electrons are bound to the atom) and are proportional to $1/n^2$. A much more sophisticated model of the atom considers the wave-like nature of the electrons from the very beginning. This uses wave mechanics or quantum mechanics. Here each electron is described by a wavefunction ψ which is a function of spatial position (x,y,z) and, in general, of time. Physically $|\psi|^2$ represents the probability of finding the electron at any point. To work out the energy of each electron, we need to solve the Schrödinger equation which, in the time-independent case, takes the form

$$\frac{\hbar^2}{2m_e}\nabla^2\psi + V(x, y, z)\,\psi = E\psi, \qquad (3)$$

where V(x,y,z) describes the potential energy function in the environment of the electron. Solution of the Schrödinger equation, under certain boundary conditions, leads to a set of solutions for the allowed wavefunctions ψ_n of the atomic electrons together with their associated energies E_n.This equation can only be solved analytically for the case of the hydrogen atom, where there is only one electron moving in the potential of a single proton, the hydrogen nucleus.

Only a certain set of electronic wavefunctions and associated energy levels fulfil this Schro¨ dinger equation. The wavefunctions may be expressed as a radial part, governing the spatial extent of the wavefunction, multiplied by a spherical harmonic function which determines the shape. The allowed wavefunctions form the electron orbitals, which we term 1s, 2s, 2p, 3s, 3p, 3d, etc. (here 1, 2, 3, . . . are alternative labels for K, L, M, . . .). These allowed wavefunctions now depend on not just one quantum number but four: *n, l, m* and *s*. These numbers may be summarised as follows:

— *n* is the principal quantum number; it is like the quantum number used for the case of Bohr shells (*n* = 1, 2, 3, . . .).

— *l* is the angular momentum quantum number; it can vary from *l* =0, 1, 2, . . ., (*n* – 1). The value of *l* governs the orbital shape of the subshell: *l* = 0 is an *s* orbital, which is spherical; *l* = 1 is a *p* orbital, which has a dumbbell shape; while *l* = 2 is a *d* orbital, which has a more complex shape such as a double dumbbell.

— *m* is the magnetic quantum number; it can vary from *m* = 0, ±1, . . . ,±l. The value of *m* governs the spatial orientation of the different orbitals within a subshell; i.e., there are three *p* orbitals (*l* = 1) p_x, p_y, and p_z corresponding to the three values of *m* which are 0, +1 and –1. In the absence of a magnetic field, all these orbitals within a particular subshell will have the same energy.

— *s* is the spin quantum number which, for an electron, can take the values ±1/2. Each (*n, l, m*) orbital can contain two electrons of opposite spin due to the Pauli exclusion principle, which states that no two electrons can have the same four quantum numbers.

— Using this identification in terms of the quantum numbers, each electron orbital in an atom therefore has a distinct combination of energy, shape and direction (x, y, z) and can contain a maximum of two electrons of opposite spin.

In an isolated atom, these localised electronic states are known as Rydberg states and may be described in terms of simple Bohr shells or as combinations of the three quantum numbers *n, l* and *m* known as electron orbitals. The Bohr shells correspond to the principal quantum numbers *n* equal to 1, 2, 3, etc. Within each of these shells, the electrons may exist in (n – 1) subshells.

The occupation of the electronic energy levels depends on the total number of electrons in the atom. In the hydrogen atom, which contains only one electron, the set of Rydberg states is almost entirely empty except for the lowest-energy 1s level which is half full. As we go to higher energies, the energy spacing between these states becomes smaller and smaller and eventually converges to a value known as the vacuum level ($n = \infty$), which corresponds to the ionisation of the inner-shell electron. Above this energy the electron is free of the atom and this is represented by a continuum of empty electronic states. In fact, the critical energy to ionize a single isolated hydrogen atom is equal to 13.61 eV and this quantity is the Rydberg constant. This description is strictly only true for hydrogen; however, other heavier atoms are found to have similar wavefunction (hydrogenic-like) solutions, which ultimately leads to the concept of the periodic table of elements, as each atom has more and more electrons which progressively fill the allowed energy levels.

The chemical properties of each atom are then principally determined by the number of valence electrons in the outermost electron shell which are relatively loosely bound and available for chemical reaction with other atomic species.

One way to picture the bonding between atoms is to use the concept of Molecular Orbital (MO) Theory. MO theory considers the electron wavefunctions of the individual atoms combining to form molecular wavefunctions. These orbitals, which are now delocalised over the whole molecule, are then occupied by all the available electrons from all the constituent atoms in the molecule.

Molecular orbitals are really only formed by the wavefunctions of the electrons in the outermost shells (the valence electrons); i.e., those which significantly overlap in space as atoms become progressively closer together; the inner electrons remain in what are essentially atomic orbitals bound to the individual atoms. A simple one-electron molecule is the H_2^+ ion, where we have to consider the interactions (both attractive and repulsive) between the single electron and two nucleii. The Born-Oppenheimer approximation regards the nuclei as fixed and this simplifies the Hamiltonian used in the Schrödinger equation for the molecular system.

For a one electron molecule, the equation can be solved mathematically, leading to a set of molecular wavefunctions which describe molecular orbitals and depend on a quantum number λ which specifies the angular momentum about the internuclear axis. Analogous to the classification of atomic orbitals (AOs) in terms of angular momentum l as s, p, d, etc., the MOs may be classified as σ, π, δ depending on the value of λ ($\lambda = 0, 1, 2$, respectively). Very simply a σ MO is formed from the overlap (actually a linear combination) of AOs parallel to the bond axis, whereas a π MO results from the overlap of AOs perpendicular to the bond axis. For the H_2^+ ion, the two lowest-energy solutions are known as $1s\sigma_g$ and $1s\sigma_u$.

Here 1s refers to the original atomic orbitals; the subscripts g and u refer to whether the MO is either symmetrical or non-symmetrical with respect to inversion about a line drawn between the nucleii (viz. an even or odd mathematical function). This is shown in Figure 1.

As can be seen the electron density is concentrated between the nuclei for the $1s\sigma_g$ MO, which is known as a bonding orbital since the energy of the molecular wavefunction is lower (i.e., more stable) than the corresponding isolated atomic wavefunctions. Conversely, the electron density is diminished between the nuclei for $1s\sigma_u$, which is known as an antibonding orbital since the energy of the molecular wavefunction is higher (i.e., less stable) than the corresponding isolated atomic wavefunction.

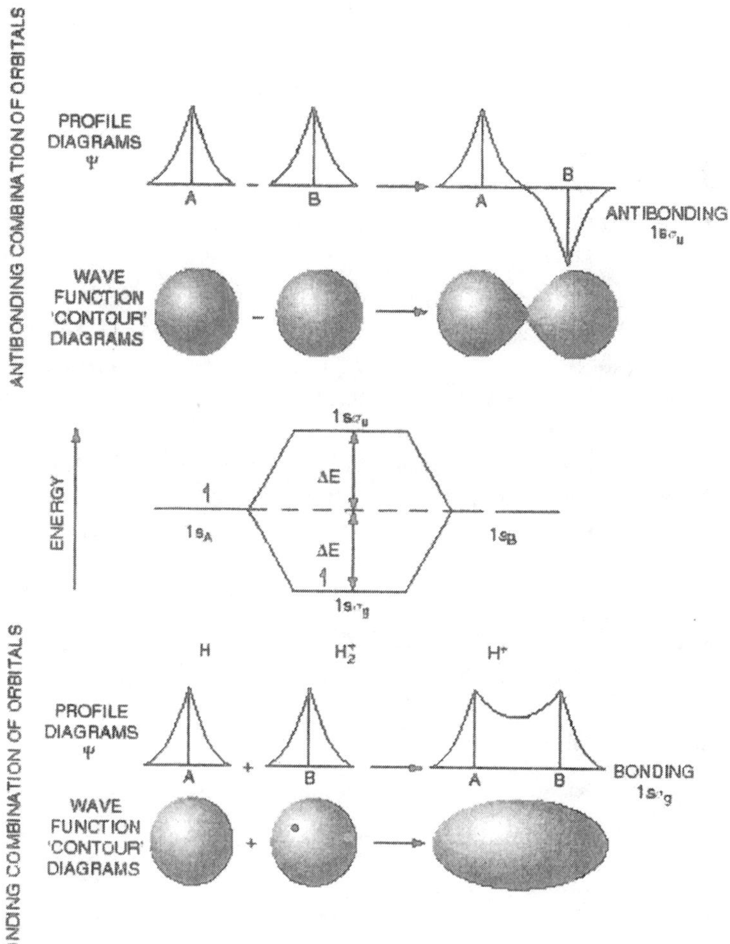

Figure 1. Molecular orbital description and energy level diagram for an H_2^+ ion

More generally, it is necessary to be able to solve the Schrödinger equation for molecules containing more than one electron. One way to do this is to use approximate sclutions similar to those obtained for the hydrogen atom, since when an electron is near a particular nucleus it will have a hydrogen-like form. Using this approach we can then construct a set of molecular orbitals from a linear combination of atomic orbitals (LCAO). Similar considerations apply to overlap of p orbitals, although now these may form both σ and π bonding and antibonding MOs.

The stability of simple diatomic molecules such as H_2, H^-_2 and He_2 depends on the relative filling of the bonding and antibonding MOs; e.g., H^-_2 contains three electrons, two of which fill the bonding MO ($1s\sigma_g$ level) while the third enters the antibonding MO ($1s\sigma_u$ level); consequently, the overall bond strength is approximately half that in H_2. Meanwhile He_2 is unstable as there are an equal number of electrons in bonding MOs as in antibonding MOs.

For polyatomic molecules such as BF_3 a greater variety of molecular orbitals can be formed. MO theory emphasises the delocalised nature of the electron distribution, so in general these MOs are extended over not just two, but all the constituent atoms. The total number of MOs is equal to the number of valence atomic orbitals used to construct them.

Giant Molecular Solids

When atoms come into close proximity with other atoms in a solid, most of the electrons remain localised and may be considered to remain associated with a particular atom. However, some outer electrons will become involved in bonding with neighbouring atoms. Briefly, the welldefined outer electron states of the atom overlap with those on neighbouring atoms and become broadened into energy bands. One convenient way of picturing this is to envisage the solid as a large molecule.

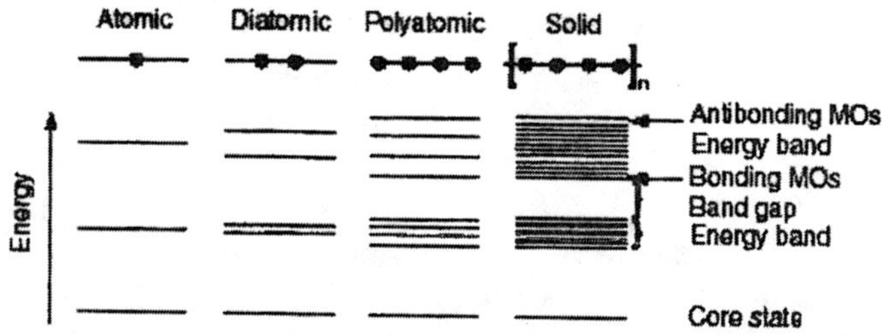

Figure 2. Electron energy level diagram for a progressively larger linear chain of atoms showing the broadening of molecular orbitals into energy bands for a one-dimensional solid

For a simple diatomic molecule, as discussed previously, the two outermost atomic orbitals (AOs) overlap to produce two molecular orbitals (MOs) which can be viewed as a linear combination of the two constituent atomic orbitals. As before, the bonding MO is formed from the in-phase overlap of the AOs and is lower in energy than the corresponding AOs, whereas the other MO, formed from the out-of-phase overlap, is higher in energy than the corresponding AOs and is termed an antibonding MO.

Progressively increasing the length of the molecular chain increases the total number of MOs, and gradually these overlap to form bands of allowed energy levels which are separated by forbidden energy regions. These band gaps may be thought of as arising from the original energy gaps between the various atomic orbitals of the isolated atoms. Note that the broadening of atomic orbitals into energy bands as the atoms are brought closer together to form a giant molecular solid can sometimes result in the overlapping of energy bands to give bands of mixed (atomic) character.

The degree to which the orbitals are concentrated at a particular energy is reflected in a quantity known as the density of states (DOS) $N(E)$, where $N(E)\ dE$ is the number of allowed energy levels per unit volume of the solid in the energy range between E and $E + dE$. As in a simple molecule, each MO energy level in the energy band can accommodate two electrons of opposite spin. The total number of electrons from all the interacting atomic orbitals in the large molecule fill this set of MOs, the highest occupied energy level being known as the Fermi level E_F. The sum of the energies of all the individual electrons in the large molecule gives the total energy of the system, which gives a measure of the stability of the atomic arrangement in terms of the system free energy.

Free Electron Model and Energy Bands

An alternative view of the electronic band structure of solids is to consider the electron waves in a periodic crystalline potential. The starting point for this approach is the Drude-Lorentz free electron model for metals. In this model a metallic solid is considered as consisting of a close packed lattice of positive cations surrounded by an electron sea or cloud formed from the ionisation of the outer shell (valence) electrons. We can then treat the valence electrons as if they were a gas inside a container and use classical kinetic gas theory. This works best for the electropositive metals of Groups I and II as well as aluminium (the so-called free electron metals) and can explain many of the fundamental properties of metals such as high electrical and thermal conductivities, optical opacity, reflectivity, ductility and alloying properties.

However, a more realistic approach is to treat the free electrons in metals quantum mechanically and consider their wave-like properties. Here the free valence electrons are assumed to be constrained within a potential well which essentially stops them from leaving the metal. The box boundary conditions require the wavefunctions to vanish at the edges of the crystal. The allowed wavefunctions given by the Schrödinger equation then correspond to certain wavelengths as shown in Figure 3. For a one-dimensional box of length L, the permitted wavelengths are $\lambda_n = 2L/n$, where $n = 1, 2, 3 \ldots$ is the quantum number of the state; the permitted wavevectors $k_n = 2\pi/\lambda$ are given by $k_n = n\lambda/L$.

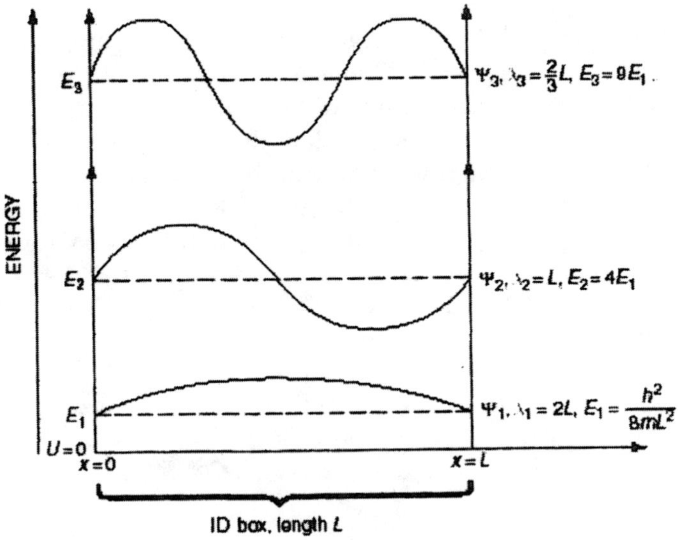

Figure 3. Energy level diagram also showing the form of some of the allowed wavefunctions for an electron confined to a one-dimensional potential well

This simple particle-in-a-box model results in a set of wavefunctions given by

$$\psi_n = (2/L)^{1/2}\sin(m\pi x/L), \tag{4}$$

where $n = 1, 2, 3 \ldots$., and for each n the corresponding energy of the electronic level is

$$E_n = \frac{n^2h^2}{8mL^2}. \tag{5}$$

E_n represents solely kinetic energy since the potential energy is assumed to be zero within the box. Thus there is a parabolic relationship between E_n and n, and therefore between E_n and k since k depends directly on n as described above. The permitted energy levels on this parabola are discrete (i.e., quantised): however in principle the size of L for most metal crystals (ranging from microns to millimetres or even centimetres) means that the separation between levels is very small compared with the thermal energy k_BT, and we can regard the energy distribution as almost continuous (quasi-continuous) so that the levels form a band of allowed energies. Note that as the electron becomes more localised (i.e., L decreases), the energy of a particular electron state (and more importantly the spacing between energy states) increases; this has important implications for bonding and also for reduced-dimensionality or quantum-confined systems.

Crystalline solids

There are three dimensions to consider the electronic properties of bulk crystalline solids. The unit cell contains atoms arranged in the characteristic configuration of the crystal, such that contiguous replication of the unit cell throughout all space is sufficient to generate the entire crystal structure. In otherwords, the crystal has translational symmetry, and the crystal structure may be generated by translations of the unit cell in all three dimensions. Translation symmetry in a periodic structure is a so-called discrete symmetry, because only certain translations—those corresponding to integer multiples of the lattice translation vectors derived from the unit cell—lead to symmetry-equivalent points.

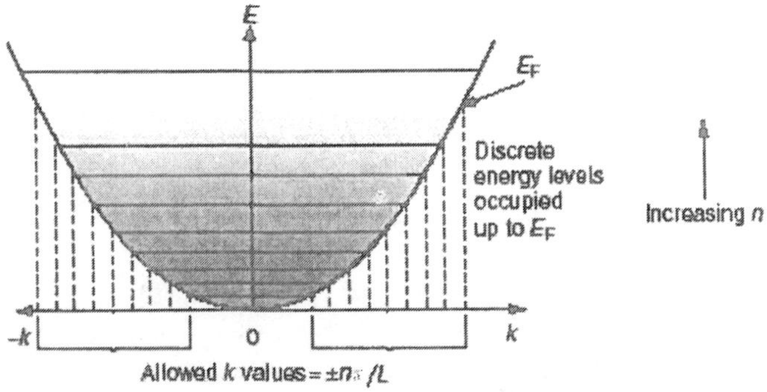

Figure 4. Schematic version of the parabolic relationship between the allowed electron wave vectors and the their energy for electrons confined to a one-dimensional potential well

Common unit cells are simple cubic, face centred cubic, body centred cubic, and the diamond structure, which comprises two interlocking facecentred cubic lattices. However, in general, the lattice spacing may be different along the different principal axes, giving rise to the orthorhombic and tetragonal unit cells, and sides of the unit cell may not necessarily be orthogonal, such as in the hexagonal unit cell.

Generally, symmetries generate conservation laws; this is known as Noether's theorem. The continuous translation symmetry of empty space generates the law of momentum conservation; the weaker discrete translation symmetry in crystals leads to a weaker quasi-conservation law for quasi- or crystal momentum.

Periodicity of crystal lattices

The three dimensional periodicity of the atomic arrangement in a crystal gives rise to a corresponding periodicity in the internal electric potential due to the ionic cores. Incorporating this periodic potential into the

Schrödinger equation results in allowed wavefunctions that are modulated by the lattice periodicity. Bloch's theorem states that these wavefunctions take the form of a plane wave (given by exp (ik.r)) multiplied by a function which has the same periodicity as the lattice; i.e.,

$$\psi(\mathbf{r}) = u_k(\mathbf{r}) \exp(i\mathbf{k}{:}\mathbf{r}); \tag{6}$$

where the function $u_k(\mathbf{r})$ has the property $u_k(\mathbf{r}+\mathbf{T}) = u_k(\mathbf{r})$, for any lattice translation vector **T**. Such wavefunctions are known as Bloch functions, and represent travelling waves passing through the crystal, but with a form modified periodically by the crystal potential due to each atomic site. For a one dimensional lattice of interatomic spacing *a*, these relationships reduce to

$$\psi(x) = u_k(x)\ u_k(x) \exp(ikx) \tag{7}$$

with $u_k(x + na) = u_k(x)$ for any integer *n*. Now, if we impose periodic boundary conditions at the ends of the chain of atoms of length $L = Na$:

$$\psi(Na) = \psi(0), \tag{8}$$

we find that

$$u_k(Na) \exp(ikNa) = u_k(0); \tag{9}$$

from which exp $(ik\ Na) = 1$, which has the solutions

$$k = \pm 2n\pi\ /\ Na = \pm 2n\pi/L \tag{10}$$

for integer *n*. This result has two important consequences: firstly, it tells us that the difference between consecutive *k* values is always $2\pi/L$, which can be interpreted as representing the volume (or properly, in this simplified one dimensional case, the length) of *k*-space occupied by each wavevector state. Applying this argument to each dimension in turn gives, for a 3D crystal, a *k*-space volume of $8\pi^3/V$ occupied by each wavevector state, where $V = L^3$ is the crystal volume. Secondly, once the upper limit on n is determined, equation (10) will also tell us how many wavevector states are contained within each energy band. However, since electrons exhibit wave-like properties, the free electrons present in the crystal also experience the same diffraction phenomena, and this has a crucial effect on the spectrum of allowed electron energies. If we consider an electron wave travelling along a one dimensional chain of atoms of spacing a, then each atom will cause reflection of the wave.

These reflections will all be constructive provided that $m\lambda = 2a$, for integer m, where λ is the electron de Broglie wavelength. When this condition is satisfied, both forward and backward travelling waves exist in the lattice, and the superposition of these creates standing waves. The standing waves correspond to electron density distributions $|\psi(x)|^2$ which have either all nodes, or all antinodes, at the lattice sites $x = a, 2a, 3a, \ldots$, and these two solutions, although having the same wavevector value, have quite different associated energies, due to the different interaction energies between the electrons and the positively charged ions.

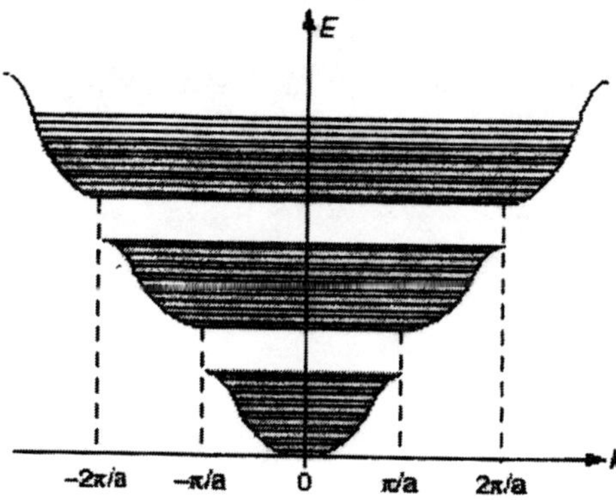

Figure 5. Schematic version of the parabolic relationship between the allowed electron wavevectors and the their energy for electrons confined

to a one-dimensional potential well containing a periodically varying potential of period a

Consequently a band gap forms in the electron dispersion curve at the corresponding values of wavevector: k = ±mp/ *a* (Figure 5). The fact that the electron waves are standing waves means that the electron group velocity

$$v_g = \frac{\partial \omega}{\partial k} = \frac{1 \partial E}{\hbar \partial k} \tag{11}$$

tends to zero at these points. This represents a fundamental difference between the behaviour of electrons in crystalline solids and that in free space, where the dispersion relationship remains purely parabolic ($E \propto k^2$) for all values of *k*. The region of *k* space which lies between any two diffraction conditions is know as a Brillouin zone: thus, in a one-dimensional crystal, the first Brillouin zone lies between $k = -\pi/a$ and $k = +\pi/a$. However, any value of *k* which lies outside the first Brillouin zone corresponds, mathematically, to an electron wave of wavelength $\lambda < 2a$.

Such a wave has too high a spatial frequency to be uniquely defined by a set of wave amplitudes which are only specified at lattice sites: an equivalent wave of wavelength $\lambda > 2a$ can always be identified. In *k*-space, this transformation is represented by the fact that any value of *k* lying outside the first Brillouin zone is equivalent to some point lying inside the first Brillouin zone, where the equivalent point is found from the relation

$$k' = k \pm 2m\pi / a \tag{12}$$

and the set of values $2m\pi/a$ are known as the reciprocal lattice vectors for the crystal. In a three dimensional crystal, the location of energy gaps in the electron dispersion is still determined from electron diffraction by the lattice planes, but the Brillouin zones are no longer simple ranges of *k*, as in 1D: rather, they are described by complex surfaces in 3D k-space, the geometry of which depends on the unit cell and atomic structure.

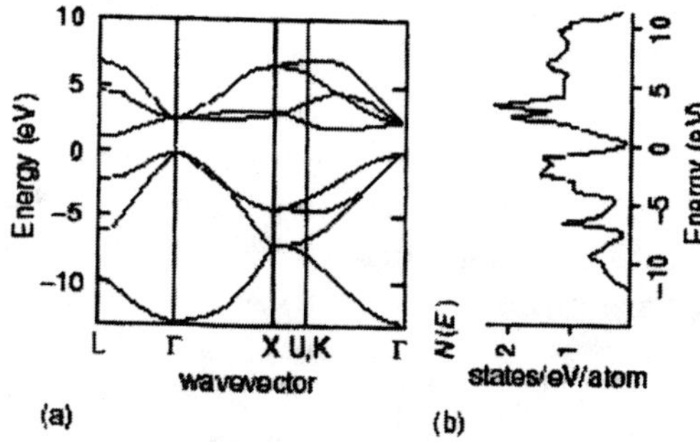

Figure 6. Electron energy band structure diagram and density of states for crystalline silicon

When the energy-wavevector relationship for such a crystal extending over multiple Brillouin zones is mapped entirely into the first Brillouin zone, this results in a large number of different energy bands and consequently the density of energy states takes on a very complex form. An example of the multiple energy bands and corresponding density of states in a real crystal is shown for the case of silicon in Figure 6.

Electronic Conduction Band

We may now observe that the series of allowed k values in equation 12 extends up to the edges of the Brillouin zone, at $k = \pm\pi/a$. Since one of these endpoints may be mapped onto the other by a reciprocal lattice vector translation, the total number of allowed k values is precisely N. Recalling that each k state may be occupied by both a spin up and a spin down electron, the total number of states available is $2N$ per energy band. In three dimensions, this result is generalised to $2N_u$ states per band, where Nu is the number of unit cells in the crystal. Now, the total number of valence electrons in the crystal is zN_u, where z is the number of valence electrons per unit cell. This leads to two very different electronic configurations in a solid. If z is even, then one energy band is completely filled, with the next band being completely empty.

The highest filled band is the valence band, and the next, empty band, is the conduction band. The electrons in the valence band cannot participate in electrical conduction, because there are no available states for them to move into consistent with the small increase in energy required by motion in response to an externally applied voltage: hence this configuration results in an insulator or, if the band gap is sufficiently small, a semiconductor. Alternatively, if z is odd, then the highest occupied energy band is only half full. In such a material, there are many vacant states immediately adjacent in energy to the highest occupied states, therefore electrical conduction occurs very efficiently and the material is a metal.

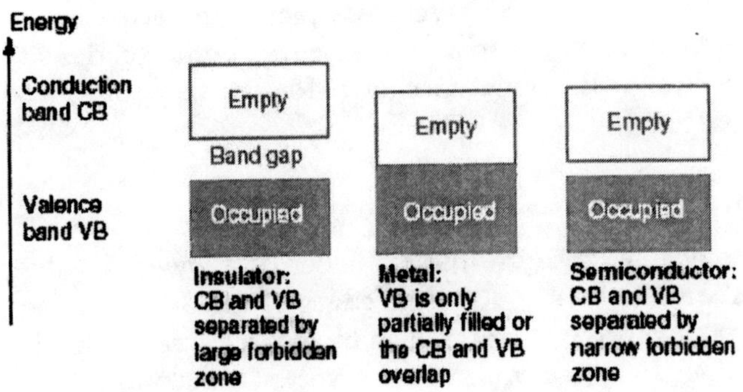

Figure 7. Electron energy band diagram for an insulator, a conductor and a semiconductor

Figure 7 shows schematic energy diagrams for insulators, metals and semiconductors respectively. There is one further, special case which gives rise to metallic behaviour: namely, when the valence band is completely full (z is even), but the valence and conduction bands overlap in energy, such that there are vacant states immediately adjacent to the top of the valence band, just as in the case of a half-filled band. Such a material is called a semi-metal.

In the same way as was defined in the molecular orbital theory of section 1.2.3, the uppermost occupied energy level in a solid is the Fermi level E_F, and the corresponding Fermi wavevector is given by $E_F = hk^2_F/2m_e$. As mentioned above, the volume of *k*-space per state is $8\pi^3/V$. Therefore, the volume of *k* space filled by *N* electrons is $4N\pi^3/V$, accounting for the fact that 2 electrons of opposite spins can occupy each wavevector state. If we equate this volume to the volume of a sphere in *k* space, of radius k_F (the Fermi sphere), we obtain the result

$$k_F = (3\pi^2 n_e)^{1/3} \tag{13}$$

where $n_e = N/V$ is the electron density, and hence

$$E_F = \hbar\,(3\pi^2 n_e)^{2/3}/2m_e. \tag{14}$$

If the Fermi sphere extends beyond the first Brillouin zone, as occurs in many metals, then the appropriate mapping back into the zone results in a Fermi surface of complex topology.

The density of states *N(E)dE* is defined such that $N_s = \int N(E)dE$ gives the total number of states per unit crystal volume in an energy band. Now, from the above argument, the number of wavevector states per unit volume of *k* space is $V/8\pi^3$. Thus, the total number of states per band may be calculated from

$$N_s = 2 \times V/8\pi^3 \int dk. \tag{15}$$

where the factor of 2 accounts for the 2 spin states per *k* value. In three dimensions, $dk = 4\pi k^2 dk$ and thus we may write

$$N_s = \frac{V}{4\pi^3} \int 4\pi k^2 \frac{dk}{dE} dE. \tag{16}$$

For parabolic bands, $E = \hbar^2 k^2/(2m_e)$ and hence $dk/dE = m_e/(\hbar^2 k)$, from which

$$N(E) = \frac{4\pi(2m_e)^{3/2} E^{1/2}}{\hbar^3}. \tag{17}$$

The dependence of the density of states on $E^{1/2}(\propto/k)$ is simply a consequence of the increased volume of phase space available at larger values of energy. The actual population of electrons as a function of energy is given by the product of the density of states and the occupation probability $f(E)$ which, for electrons or holes, is given by the Fermi Dirac function

$$f(E) = \frac{1}{\exp((E - E_F)/k_B T) + 1}, \tag{18}$$

from which we may observe that the Fermi energy corresponds to the energy at which the occupation probability is exactly one half.

In the zero temperature limit, $f(E) = 1$ for all $E < E_F$, and $f(E) = 0$ for all $E > E_F$ in otherwords, all electron states below the Fermi energy are filled, and all those above E_F are empty, as previously described; the electron population at any energy $E < E_F$ is then given just by $N(E)$. At non-zero temperatures $f(E)$ describes the fact that some electrons are thermally excited from states just below E_F to states just above E_F, and the sharpness of the cut-off of $N(E)$ at E_F decreases with increasing temperature. Both zero temperature and non-zero temperature cases are shown in Figure 8.

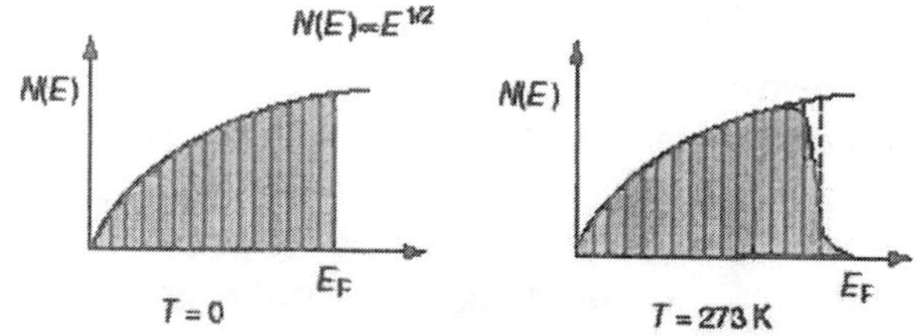

Figure 8. The density of electron states for free electrons and the occupation of electron energy levels (shaded region) at zero and room temperature

In addition to the total DOS, which has already been mentioned, it is possible to project the DOS onto a particular atomic site in the unit cell and determine the so-called local DOS; this is the contribution of that particular atomic site to the overall electronic structure. If a unit cell contains a particular type of atom in two distinct crystallographic environments, then the local DOS will be correspondingly different. Similar projections may be performed in terms of the angular momentum symmetry (i.e., the s, p, d or f atomic character of the DOS). Until now we have been concerned with crystalline systems. However, it is also possible to consider the DOS of an amorphous material; here the DOS is primarily determined by the short-range order in the material; i.e., the nearest neighbours. An alternative approach is to represent the amorphous solid by a very large unit cell with a large number of slightly different atomic environments.

Influence of Nanometre Length Scale

The small length scales present in nanoscale systems directly influence the energy band structure and can lead indirectly to changes in the associated atomic structure. Such effects are generally termed quantum confinement.

Changing Total Energy

In the free electron model, it is clear that the energies of the electronic states depend on $1/L^2$ where L is the dimension of the system in that particular direction; the spacing between successive energy levels also varies as $1/L^2$. This behaviour is also clear from the description of a solid as a giant molecule: as the number of atoms in the molecule increases, the MOs gradually move closer together. Thus if the number of atoms in a system, hence the length scale, is substantially different to that in a normal bulk material, the energies and energy separations of the individual electronic states will be very different. Although in principle the Fermi level would not be expected to change since the free electron density N/V should remain constant, there may be associated modifications in structure which will change this quantity. Furthermore, as the system size decreases, the allowed energy bands become substantially narrower than in an infinite solid.

The normal collective (i.e., delocalised) electronic properties of a solid become severely distorted and the electrons in a reduced-dimensional system tend to behave more like the 'particle in a box' description; this is the phenomenon of quantum confinement. In otherwords, the electronic states are more like those found in localised molecular bonds rather than those in a macroscopic solid. The main effect of these alterations to the bulk electronic structure is to change the total energy and hence, ignoring entropy considerations, the thermodynamic stability of the reduced length scale system relative to that of a normal bulk crystal. This can have a number of important implications. It may change the most energetically stable form of a particular material; for example, small nanoparticles or nanodimensional layers may adopt a different crystal structure from that of the normal bulk material.

Some metals which normally adopt a hexagonal close-packed atomic arrangement have been reported to adopt a face-centred cubic structure in confined systems such as metallic multilayers. If a different crystallographic structure is adopted below some particular critical length scale, then this arises from the corresponding change in the electronic density of states which often results in a reduced total energy for the system. Reduction of system size may change the chemical reactivity, which will be a function of the structure and occupation of the outermost electronic energy levels.

Correspondingly, physical properties such as electrical, thermal, optical and magnetic characteristics, which also depend on the arrangement of the outermost electronic energy levels, may be changed. For example, metallic systems can undergo metal-insulator transitions as the system size decreases, resulting from the formation of a forbidden energy band gap. Other properties such as mechanical strength which, to a first approximation, depends on the change in electronic structure as a function of applied stress and hence interatomic spacing, may also be affected. Transport properties may also change in that they may now exhibit a quantised rather than continuous behaviour, owing to the changing nature and separation of the electron energy levels.

A related viewpoint for understanding the changes observed in systems of reduced dimension is to consider the proportion of atoms which are in contact with either a free surface, as in the case of an isolated nanoparticle, or an internal interface, such as a grain boundary in a nanocrystalline solid. Both the surface area to volume ratio (*S/V*) and the specific surface area (m^2g^{-1}) of a system are inversely proportional to particle size and both increase drastically for particles less than 100 nm in diameter. For isolated spherical particles of radius *r* and density ρ, the surface area per unit mass of material is equal to $4\pi r^2/(4/3\lambda r^3\rho) = 3/r\rho$. For 2 nm diameter spherical particles of typical densities, the specific surface area (SSA) can approach 500m² g⁻. However, for particles in contact this value will be reduced by up to approximately half. This large surface area term will have important implications for the total energy of the system.

As discussed above this may lead to the stabilisation of metastable structures in nanometre-sized systems, which are different from the normal bulk structure or, alternatively, may induce a simple relaxation (expansion or contraction) of the normal crystalline lattice which could in turn alter other material properties. If an atom is located at a surface then it is clear that the number of nearest-neighbour atoms are reduced, giving rise to differences in bonding (leading to the well-known phenomenon of surface tension or surface energy) and electronic structure. In a small isolated nanoparticle, a large proportion of the total number of atoms will be present either at or near the free surface.

For instance, in a 5nm particle approximately 30-50% of the atoms are influenced by the surface, compared with approximately a few percent for a 100nm particle. Similar arguments apply to nanocrystalline materials, where a large proportion of atoms will be either at or near grain boundaries. Such structural differences in reduced-dimensional systems would be expected to lead to very different properties from the bulk.

Another important consideration for nanostructures concerns the number of atomic vacancies nv which exist in thermal equilibrium in a nanostructure. Vacancies are point defects in the crystalline structure of a solid and may control many physical properties in materials such as conductivity and reactivity. In microcrystalline solids at temperatures above 0K, vacancies invariably exist in thermal equilibrium. In the simple case of metals with one type of vacancy, the number of vacancies in a crystal consisting of N atom sites is approximated by an Arrhenius-type expression

$$n_v = N \exp(-Q_f / RT), \tag{19}$$

where T is the absolute temperature, R is the gas constant and Q_f is the energy required to form one mole of vacancies. Q_f is given by the relationship $Q_f = N_A q_f$, where N_A is the Avogadro number and q_f is the activation energy for the formation of one vacancy. However, the value of qf is not well defined but is generally estimated to be the energy required to remove an atom from the bulk interior of a crystal to its surface. As a rough approximation, a surface atom is bonded to half the number of atoms compared with an interior atom, so q_f represents half the bonding energy per atom. Since the melting temperature T_m of a metal is also a measure of the bond energy, then q_f is expected to be a near linear function of T_m.

From a continuum model, Q_f may be estimated from the latent heat of vaporisation, since on leaving the surface an atom breaks the remaining (half) bonds. In practice it is found that the latent heat of vaporisation is considerably higher than Q_f. Alternatively, Q_f may be estimated from the surface energy per unit area. Given that one atom occupies an area b^2, the number of atoms per unit area is equal to $1/b^2$ and the surface energy σ is therefore q_f/b^2. Surface energies depend on melting temperature and vary within the range 1:1 Jm^{-2} (for aluminium) to 2:8 Jm^{-2} (for tungsten). Taking an average value of σ as 2:2 Jm^{-2} and b = 2:5 $\times 10^{-10}$ m, we may calculate $Q_f = N_A \sigma b^2$ as 83×10^3 J mol^{-1}, which is close to the accepted value of 90 kJmol. Furthermore, the value of Q_f may be modified for nanoparticles through the influence of the surface energy term, σ, which is related to the internal pressure, P, by the simple relationship $P = 4\sigma/d$, where d is the diameter of the nanoparticle. The effect of P is to require an additional energy term, q_n, for the formation of a vacancy, which is approximately given by Pb^3. Again taking σ as 2:2 Jm^{-2}, we may calculate this additional energy per mole $Q_n = N_A q_n$ as 8:3 × 10^3 J mol^{-1} for a 10 nm diameter nanoparticle.

This term is only approximately 10% of Q_f and rapidly decreases for larger particle sizes. Thus we may conclude that the effect of the surface energy (internal pressure) factor on the vacancy concentration will be small. Additionally, the internal pressure *P* results in an elastic, compressive volume strain, and hence linear strain, ", given approximately as

$$\varepsilon = P / 3E = 4\sigma / 3dE \qquad (20)$$

where *E* is the Young's modulus. This expression suggests that the linear strain will be inversely proportional to particle size and that there will be a decrease in lattice parameter or interatomic spacing for small nanoparticles.

Finally, substituting a value of Q_f = 90 × 10^3 J mol^{-1} into the Arrhenius expression (19) for the vacancy concentration, we obtain values for the ratio n_v/N of 2:4 × 10^{-16} (at 300 K), 6:5 × 10^{-7} (at 600K) and 4:8 × 10^{-4} (at 1000 K), illustrating the exponential increase in vacancy concentration with temperature. Now consider a spherical nanoparticle, say 50 nm in diameter, and in which each atom occupies a volume b^3. Again taking b = 0:25nm, there are a total of 4:2 × 10^6 atoms in the particle, which implies that $n_v << 1$, except at very high temperatures.

Therefore nanocrystals are predicted to be essentially vacancy-free; their small size precludes any significant vacancy concentration. This simple result also has important consequences for all thermomechanical properties and processes (such as creep and precipitation) which are based on the presence and migration of vacancies in the lattice.

Dislocations in cystalline structure

Planar defects, such as dislocations, in the crystalline

structure of a solid are extremely important in determining the mechanical properties of a material. It is expected that dislocations would have a less dominant role to play in the description of the properties of nanocrystals than in the description of the properties of microcrystals, owing to the dominance of crystal surfaces and interfaces.

The free energy of a dislocation is made up of a number of terms: (i) the core energy (within a radius of about three lattice planes from the dislocation core); (ii) the elastic strain energy outside the core and extending to the boundaries of the crystal, and (iii) the free energy arising from the entropy contributions. In microcrystals the first and second terms increase the free energy and are by far the most dominant terms. Hence dislocations, unlike vacancies, do not exist in thermal equilibrium.

The core energy is expected to be independent of grain size. Estimates are close to 1 eV per lattice plane which, for an interplanar spacing b of 0.25 nm, translates to a value of about 6:5 $\times$ 10^{-10} Jm^{-1}. The elastic strain energy per unit length for an edge dislocation is given by

$$E = \frac{Gb^2}{4\pi(1-v)} \times \ln\left(\frac{r_1}{r_0}\right) \tag{21}$$

where G is the bulk modulus, r_0 is the core radius, r_1 is the crystal radius and v is Poisson's ratio. v is typically around 1/3 for a crystalline sample. The expression for a screw dislocation omits the (1 - v) term, giving an energy about 2/3 that of an edge dislocation. For G = 40 $\times$ 10^9 Pa, the constant term $Gb^2/4\pi(1 - v)$ has a value of 3 $\times$ 10^{-10} Jm^{-1}. The grain size dependence is given in the ln (r_1/r_0) term, which for grain size $(2r_1)$ values of 10, 50, 1000 and 10 000 nm increases as 3, 4.6, 7.6 and 9.9 respectively. Hence it can be seen that the elastic strain energy of dislocations in nanoparticles and nanograined materials is about one-third of that in microcrystals and that, for a 10nm grain size the core energy is comparable with the elastic strain energy. In comparison, the core energy is about one-tenth of the elastic strain energy for a microcrystal. This reduction in the elastic strain energy of dislocations in nanocrystals has important consequences.

The forces on dislocations due to externally applied stresses are reduced by a factor of about three and the interactive forces between dislocations are reduced by a factor of about 10. Hence recovery rates and the annealing out of dislocations to free surfaces are expected to be reduced. For a dislocation near the surface of a semi-infinite solid, the stress towards the surface is given by the interaction of the stress field of an image dislocation at an equal distance on the opposite side.

Since nanocrystals do not approximate to semi-infinite solids, such image stresses will operate across all surfaces and the net effect, together with the reduced elastic strain energy, results in dislocations that are relatively immobile. Finally, the entropy contributions to the free energy arise as a result of (i) configurational entropy (i.e., the dislocation can be arranged in a variety of ways), (ii) a further contribution if the dislocation is assumed to be perfectly flexible, and (iii) the effect of the dislocation on the thermal vibrations of the crystal.

Factors (ii) and (iii) are independent of crystal size and their values can be estimated to be $2k_BT$ and $3k_BT$, respectively, per atomic plane. Assuming a temperature of 300 K, these values correspond to about 3×10^{-11} Jm^{-1}, considerably less than the core and elastic strain energy terms. The configurational entropy contribution to the free energy is given by

$$E = \frac{bk_BT}{L} \ln\left(\frac{L^2}{b^2}\right) \tag{22}$$

per atom plane, where L is the length of the dislocation. At 300K this gives values of $3{:}0 \times 10^{-12}$, $5{:}7 \times 10^{-14}$ and $7{:}6 \times$

10^{-15} Jm^{-1} for L = 10, 1000 and 10 000 nm, respectively. These values are again much smaller than the core and elastic strain energy terms and hence it may be concluded that dislocations in nanocrystals, as with microcrystals, do not exist as thermodynamically stable lattice defects.

A further effect, previously mentioned, is the apparent stability of metastable structures in small nanoparticles and clusters, such that all traces of the usual bulk atomic arrangement become lost. Metallic nanoparticles, such as gold, are known to adopt polyhedral shapes such as cube-octahedra, multiply twinned icosahedra and multiply twinned decahedra (Figure 9).

These nanoparticles may be regarded as multiply twinned crystalline particles (MTPs) in which the shapes can be understood in terms of the surface energies of various crystallographic planes, the growth rates along various crystallographic directions and the energy required for the formation of defects such as twin boundaries. However, there is compelling evidence that such particles are not crystals but are quasiperiodic crystals or crystalloids. These icosahedral and decahedral quasicrystals form the basis for further growth of the nanocluster, up until a size where they will switch into more regular crystalline packing arrangements.

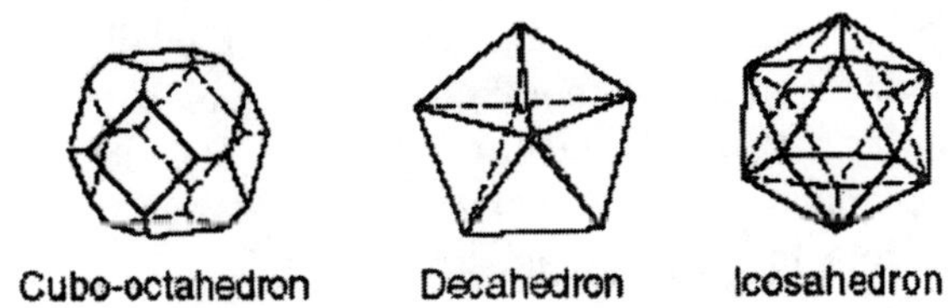

Figure 9. Geometrical shapes of cubo-octahedral particles and multiply twinned decahedral and icosahedral particles

Crystalline solids are distinct from amorphous solids in that they possess long-range periodic order and the patterns and symmetries which occur correspond to those of the 230 space groups. Quasiperiodic crystals do not possess such long-range periodic order and are distinct in that they exhibit fivefold symmetry, which is forbidden in the 230 space groups. In the cubic close-packed and hexagonal close-packed structures, exhibited by many metals, each atom is coordinated by 12 neighbouring atoms.

All of the coordinating atoms are in contact, although not evenly distributed around the central atom. However, there is an alternative arrangement in which each coordinating atom is situated at the apex of an icosahedron and in contact only with the central atom. If however we relax this 'rigid atomic sphere' model and allow the central atom to reduce in diameter by 10%, the coordinating atoms come into contact and the body now has the shape and symmetry of a regular icosahedron with point group symmetry 235, indicating the presence of 30 twofold, 20 threefold and 12 fivefold axes of symmetry. This geometry represents the nucleus of a quasiperiodic crystal which may grow in the forms of icosahedra or pentagonal dodecahedra.

These are dual solids with identical symmetry, the apices of one being replaced by the faces of the other. Such quasiperiodic crystals are known to exist in an increasing number of aluminium-based alloys and may be stable up to microcrystalline sizes. It should be noted that their symmetry is precisely the same as that of the fullerenes C_{20} (dodecahedrene with 12 pentagonal faces of a pentagonal dodecahedron, but unstable) and C_{60}. Hence, like the fullerenes, quasiperiodic crystals are expected to have an important role to play in nanostructures.

The size-related instability characteristics of quasiperiodic crystals are not well understood. A frequently observed process appears to be that of multiple twinning, such crystals being distinguished from quasiperiodic crystals by their electron diffraction patterns. Here the five triangular faces of the fivefold symmetric icosahedron can be mimicked by five twin-related tetrahedra through relatively small atomic movements.

The large increase in surface energy and the change in interatomic spacing as a function of nanoparticle size mentioned above have a marked effect on material properties. For instance, the melting point of gold particles, which is really a bulk thermodynamic characteristic, has been observed to decrease rapidly for particle sizes less than 10 nm, as shown in Figure 10. There is evidence that for metallic nanocrystals embedded in a continuous matrix the opposite behaviour is true; i.e., smaller particles have higher melting points.

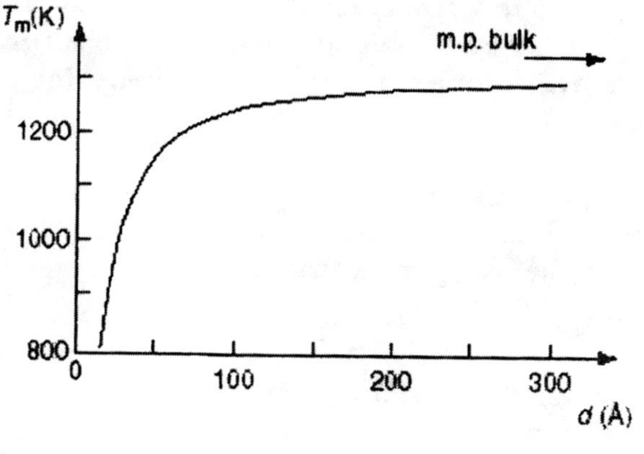

Figure 10. Schematic diagram of the variation in melting point of gold nanoparticles as a function of particle size

Chemical properties

The change in structure as a function of particle size is intrinsically linked to the changes in electronic properties. The ionisation potential (the energy required to remove an electron) is generally higher for small atomic clusters than for the corresponding bulk material. Furthermore, the ionisation potential exhibits marked fluctuations as a function of cluster size. Such effects appear to be linked to chemical reactivity, such as the reaction of Fen clusters with hydrogen gas. Nanoscale structures such as nanoparticles and nanolayers have very high surface area to volume ratios and potentially different crystallographic structures which may lead to a radical alteration in chemical reactivity.

Catalysis using finely divided nanoscale systems can increase the rate, selectivity and efficiency of chemical reactions such as combustion or synthesis whilst simultaneously significantly reducing waste and pollution. Gold nanoparticles smaller than about 5nm in diameter are known to adopt icosahedral structures rather than the normal face centred cubic arrangement. This structural change is accompanied by an extraordinary increase in catalytic activity. Furthermore, nanoscale catalytic supports with controlled pore sizes can select the products and reactants of chemical reactions based on their physical size and thus ease of transport to and from internal reaction sites within the nanoporous structure.

Additionally, nanoparticles often exhibit new chemistries as distinct from their larger particulate counterparts; for example, many new medicines are insoluble in water when in the form of micron-sized particles but will dissolve easily when in a nanostructured form.

Biological systems contain many examples of nanophase materials and nanoscale systems. Biomineralisation of nanocrystallites in a protein matrix is highly important for the formation of bone and teeth, and is also used for chemical storage and transport mechanisms within organs. Biomineralisation involves the operation of delicate biological control mechanisms to produce materials with well-defined characteristics such as particle size, crystallographic structure, morphology and architecture.

Generally complex biological molecules such as DNA have the ability to undergo highly controlled and hierarchical self-assembly, which makes them ideal for the assembling of nanosised building blocks. Methods for altering and controlling these interactions and building nanoscale building blocks and assembling nanoscale architectures are also discussed.

Biological cells have dimensions within the range 1-10mm and contain many examples of extremely complex nano-assemblies, including molecular motors, which are complexes embedded within membranes that are powered by natural biochemical processes. Generally, naturally occurring biological nanomaterials have been refined by evolutionary processes over a long timescale and are therefore highly optimised.

Methods of Fabrication

Nanostructures can be made in numerous ways. A broad classification divides methods into either those which build from the bottom up, atom by atom, or those which construct from the top down using processes that involve the removal or reformation of atoms to create the desired structure. The two approaches are schematically represented in Figure 11. In the bottom-up approach, atoms, molecules and even nanoparticles themselves can be used as the building blocks for the creation of complex nanostructures; the useful size of the building blocks depends on the properties to be engineered. By altering the size of the building blocks, controlling their surface and internal chemistry, and then controlling their organisation and assembly, it is possible to engineer properties and functionalities of the overall nanostructured solid or system. These processes are essentially highly controlled, complex chemical syntheses.

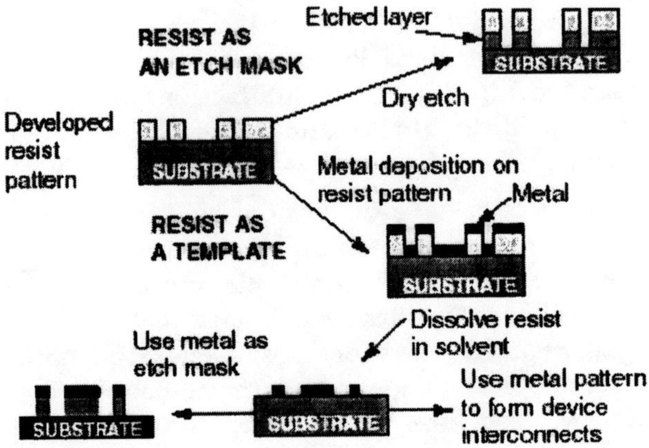

Figure 11. Schematic representation of the top-down and bottom-up processes and their relationship to biological processes and structures

On the other hand, top-down approaches are inherently simpler and rely either on the removal or division of bulk material, or on the miniaturisation of bulk fabrication processes to produce the desired structure with the appropriate properties. When controlled, both top-down and bottom-up methods may be viewed as essentially different forms of microstructural engineering. As shown in Figure 11, in terms of scale, biological processes are essentially intermediate between top-down and bottom-up processes, however in reality they usually constitute complex bottom-up processes.

Lithographic processes

Conventional lithographic processes are akin to the emulsion-based photographic process and can be used to create nanostructures by the formation of a pattern on a substrate via the creation of a resist on the substrate surface. One lithographic method uses either visible or ultraviolet (UV) light, X-rays, electrons or ions to project an image containing the desired pattern onto a surface coated with a photoresist material; this method requires the prior fabrication of an absorbing mask through which the parallel radiation passes before shadowing onto the photoresist.

Alternatively, primary patterning (or direct writing) of the resist is possible using a focused electron, ion or possibly X-ray beam; here either the focused beam or the resist itself is scanned according to the desired pattern design. These different techniques are generally termed photolithography, X-ray lithography, electron beam lithography or ion beam lithography depending on the radiation employed. The resist material, typically a polymer, metal halide or metal oxide, is chemically changed during irradiation, often altering the solubility or composition of the exposed resist. In a positive resist, irradiated areas are dissolved as they are more soluble than unexposed material, whilst in a negative resist irradiated areas are insoluble.

The pattern transfer processes which utilise the patterned resist may be divided into solution-based wet chemical etching procedures, dry etching in a reactive plasma, doping using ion implantation techniques, or thin film deposition. Dry etching is the collective term for a range of techniques such as reactive ion etching (RIE) and chemically assisted ion beam etching (CAIBE), which are used extensively for highresolution pattern transfer; both methods produce, either directly or indirectly, reactive ion species which combine with the elements in the substrate material to form volatile reaction products which are removed into a vacuum system.

Lithographic techniques are very heavily used in the semiconductor processing industry for the fabrication of integrated circuits, optoelectronic components, displays and electronic packaging. The important considerations here are not only the uniformity and reproducibility of the fabrication process, but also the time required to pattern a given area of a planar device, which is summarised in terms of the areal throughput (in microns per hour). A further consideration for nanostructures, particularly for the production of integrated circuits and microelectro-mechanical systems (MEMS), is the ultimate resolution of the lithographic technique.

Fundamentally, the wavelength of the radiation used in the lithographic process determines the detail in the resist and hence the final planar nanostructure; additional considerations may involve the limitations of the projection optics and the nature of the interaction of the radiation with the resist material.

Typically the resolution ranges from a few hundred nanometres for optical techniques to tens of nanometres for electron beam techniques. Phenomenologically, throughput and resolution of lithographic techniques broadly follow a power-law dependence; the resolution is approximately equal to $23A^{0.2}$, where A is the areal throughput. Optical, UV and X-ray lithography are fast, parallel exposure techniques capable of micron and submicron resolution, whereas electron and ion beam methods provide nanometre resolution, but are, at present, considerably slower as they involve serial exposure using scanned focused probes controlled by computer-aided design (CAD) software.

Parallel projection electron beam systems (such as the Scalpel method) and X-ray proximity printing methods are currently being developed for high-resolution, large-throughput mask fabrication. A further area which may be developed in the future is the use of massive arrays of individually controllable atomic force microscope (AFM) or scanning tunnelling microscope (STM) tips for parallel atom manipulation on planar substrates; these are essentially MEMS devices integrated with microprocessor control for the direct patterning of resists (e.g., self-assembled monolayers) or even substrates at reasonable throughputs.

Soft lithography techniques pattern a resist by physically deforming (or embossing) the resist shape with a mould or stamp, rather than by modifying the resist chemical structures with radiation as in conventional lithography. Additionally the stamp may be coated with a chemical that reacts with the resist solely at the edges of the stamp. These methods circumvent many of the resolution limitations inherent in conventional lithographic processes that arise due to the diffraction limit of the radiation, the projection or scanning optics, the scattering process and the chemistry within the resist material. Ultimately, nanoimprinting should represent a cheaper process for mass production.

Currently, these soft lithography techniques car produce patterned structures in the range 10 nm and above. One of the main limitations on resolution arises from plastic flow of the polymeric materials involved. Master moulds may be fabricated using either conventional lithographic techniques, micromachining or naturally occurring surface relief on the substrate materials.

Once the master mould is formed, a low molecular weight prepolymer is poured and then cured to yield a polydimethylsiloxane (PDMS) elastomeric 'stamp' which may simply be peeled off the master. The relief features that result on the bottom of the PDMS stamp are an inverted replica of those on the master. They may be inked with a molecule of interest (typically by wiping a Q-tip, dampened with the molecule of interest, across their surfaces) and simply placed against a substrate, transferring molecules in a pattern that reflects the features of the stamp. Elastomeric PDMS stamps have been used to impressive effect as conformal contact masks for phase-shifting photolithography, facilitating the fabrication of structures as small as 50 nm in photoresist. Channels formed in PDMS stamps have also been used to mould three-dimensional structures, a process known as micromoulding in capillaries (MIMIC).

Another variation of this technique is lithographically induced self-assembly, in which a mask is used to induce and control, via electrostatic and hydrodynamic instabilities, the formation of periodic supramolecular pillar arrays on a thin polymer melt initially deposited flat onto a substrate. Alternatively, microcontact printing of self-assembled monolayers (SAMs) can be achieved by transfer of a SAM 'ink' onto a substrate surface from a patterned 'stamp' possessing elastomeric properties so as to achieve intimate contact with the surface. A related, and rapidly emerging technique is direct inkjet printing of (currently) micron-scale structures onto surfaces.

Lithographic techniques, whilst being a parallel batch processes, essentially consist of a two-dimensional chemical or mechanical patterning of the surface of a material. More intricate three-dimensional patterning of a material can be achieved by techniques analogous to more conventional machining. Currently the resolution limits of conventional machining are of the order of 5 mm, however, in recent years focused ion beams (FIBs) and high-intensity lasers have been used to directly pattern or shape materials at micron and submicron levels. The length scale achievable in such direct sculpture of materials is not only determined by the spot size and power (or current) density of the radiation employed but also by the nature of the material itself and the removal process.

Modern FIB machines have very accurate sample manipulation stages and are not only capable of material removal, via sputtering of surface substrate atoms by the computer-controlled scanning of focused ion beams produced by liquid metal gallium ion guns, but also of material deposition using the interaction of the ion beam with reactive gases from localised microinjectors. Modern dual-beam FIBs can simultaneously view the etching or deposition procedures using a separate scanning electron microscope column. As well as inorganic materials, there is increasing interest in FIB patterning of biological structures. Thin sample preparation of specimens for transmission electron microscopy is routinely undertaken in a FIB.

Bottom-up processes effectively encompass chemical synthesis and/or the highly controlled deposition and growth of materials. Chemical synthesis may be carried out in either the solid, liquid or gaseous state. Solid-state synthesis usually involves an iterative procedure of bringing solid reaction precursors into intimate contact by mixing and grinding and then promoting atomic diffusion processes via heat treatment at high temperatures to form a reaction product. Such elevated temperatures often lead to rapid grain growth and ultimately a final product with a relatively large grain size unless grain growth inhibitors are present. Consequently, true nanoscale systems are difficult to obtain via solid-state routes, apart from perhaps the controlled growth of second phase nanoscale precipitates within a primary matrix such as in precipitation-strengthened steels, aluminium alloys or other nanocomposite systems. Here we focus primarily on liquid and gas phase fabrication routes.

Diffusion of matter in the liquid and gas phases is typically many orders of magnitude greater than in the solid phase; therefore these synthetic methods can be implemented at much lower temperatures, thus inhibiting unwanted grain growth and so resulting in true nanoscale systems.

Vapour phase deposition can be used to fabricate thin films, multilayers, nanotubes, nanofilaments or nanometre-sized particles. The general techniques can be classified broadly as either physical vapour deposition (PVD) or chemical vapour deposition (CVD). PVD involves the conversion of solid material into a gaseous phase by physical processes; this material is then cooled and redeposited on a substrate with perhaps some modification, such as reaction with a gas.

Examples of PVD conversion processes include thermal evaporation (such as resistive or electron beam heating or even flame synthesis), laser ablation or pulsed laser deposition (where a short nanosecond pulse from a laser is focused onto the surface of a bulk target), spark erosion and sputtering (the removal of a target material by bombardment with atoms or ions). One example of a PVD technique is vapour phase expansion, which relies on the expansion of a high-pressure vapour phase through a jet into a low-pressure ambient background to produce supersaturation of the vapour.

Flow rates can approach supersonic speeds and the process can lead to the nucleation of extremely small clusters, ranging from a few atoms upwards, which can be analysed using mass spectrometry. Although the quantities of such clusters are low, these can be of great use for studying the physics of small, usually metallic, clusters, which often are composed of magic numbers of atoms due to their very stable geometric and electronic configurations.

The metal vapour cools through collisions with the inert gas species, becomes supersaturated and then nucleates homogeneously; the particle size is usually in the range 1-100nm and can be controlled by varying the inert gas pressure. Particles can be collected on a cold finger cooled by liquid nitrogen, scraped off and compacted to produce a dense nanomaterial.

CVD involves the reaction or thermal decomposition of gas phase species at elevated temperatures (typically 500-1000°C) and subsequent deposition onto a substrate. A simple example is aerosol spray pyrolysis involving aqueous metal salts which are sprayed as a fine mist, dried and then passed into a hot flow tube where pyrolysis converts the salts to the final products. Since materials are mixed in solution, homogeneous mixing at the atomic level is possible and pyrolysis at relatively low temperatures provides particles in the size range 5-500 nm.

Several CVD processes employ catalysts to enhance the rates of certain chemical reactions. When the catalyst is in the form of a nanometre-sized dispersion of particles, this can often provide a templating effect; for example, in the prodution of carbon nanotubes using the decompostion of ethyne or ethene with hydrogen, Fe-, Co- or Ni-based catalysts are employed. The size and size distribution of the catalyst particles may determine the internal diameter of the nanotube.

The use of plasmas (i.e., ionised gases) during vapour deposition allows access to substantially different chemical and physical processes and also higher-purity final materials relative to the conventional PVD and CVD processes described above. There are several different types of plasma deposition reactor for plasma-assisted PVD.

DC glow discharge

DCglow discharge involves the ionisation of gas atoms by electrons emitted from a heated filament. The gas ions in the plasma are then accelerated to produce a directed ion beam. If the gas is a reactive precursor gas, this ion beamis used to deposit directly onto a substrate; alternatively an inert gas may be used and the ion beam allowed to strike a target material which sputters neutral atoms onto a neighbouring substrate (Figure 12).

Magnetron sputtering

Magnetron sputtering involves the creation of a plasma by the application of a large DC potential between two parallel plates. A static magnetic field is applied near a sputtering target and confines the plasma to the vicinity of the target.

Ions from the high-density plasma sputter material, predominantly in the form of neutral atoms, from the target onto a substrate.

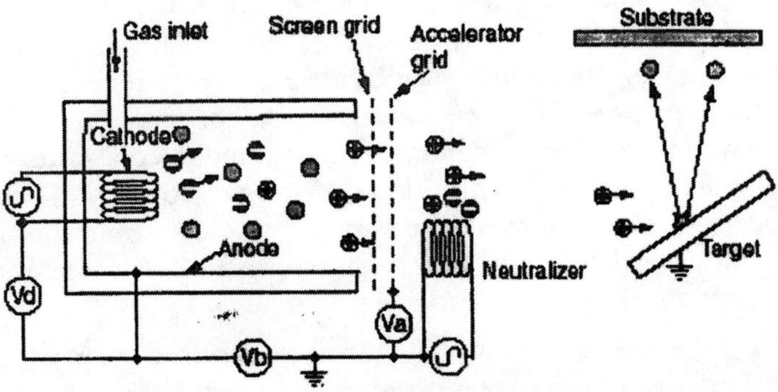

Figure 12. Schematic diagram of a DC glow discharge apparatus in which gas atoms are ionised by an electron filament and either deposit on a substrate or cause sputtering of a target

A further benefit of the magnetic field is that it prevents secondary electrons produced by the target from impinging on the substrate and causing heating or damage. The deposition rates produced by magnetrons are high enough (~1 μm/min) to be industrially viable; multiple targets can be rotated so as to produce a multilayered coating on the substrate.

MBE and MOVPE

A molecular beam epitaxy (MBE) machine is essentially an ultra-high-precision, ultraclean evaporator, combined with a set of in-situ tools, such as RHEED and Auger spectroscopy, for characterisation of the deposited layers during growth. The reactor consists of an ultra-high-vacuum chamber of approximately 1.5m diameter. MBE is a growth technique in which epitaxial, single atomic layers are grown on a heated substrate under UHV conditions, using either atomic or molecular beams evaporated from effusion sources with openings directed towards a heated substrate usually consisting of a thin wafer cut from a bulk single crystal. The sources can be either solid or gaseous and an MBE machine will typically have an array of multiple sources, which can be shuttered to allow layered, alternating heterostructures to be produced. Semiconductor quantum wells, superlattices and quantum wires and metallic or magnetic multilayers for spin valve structures are deposited using this technique. Standard MBE uses elements in a very pure form as solid sources contained within a number of Knudsen cells.

In operation the cells are heated to the temperature at which the elements evaporate, producing beams of atoms which leave the cells. The beams intersect at the substrate and deposit the appropriate semiconductor, atomic layer by atomic layer. The substrate is rotated to ensure even growth over its surface. By operating mechanical shutters in front of the cells, it is possible to control which semiconductor or metal is deposited. Opening the Ga and As cell shutters results in the growth of GaAs. Shutting the Ga cell and opening the Al cell switches the growth to AlAs.

As the shutters can be switched rapidly, in comparison to the rate at which material is deposited, it is possible to grow very thin layers exhibiting very sharp interfaces. Other effusion cells contain elements required for doping, and it is possible to monitor the growth by observing the electron diffraction pattern produced by the surface. MBE can also be performed using gaseous sources, and this is often termed chemical beam epitaxy (CBE). When the sources are metallorganic compounds, the process is known as metallorganic MBE (MOMBE). The second main epitaxial technique is metallorganic vapour phase epitaxy (MOVPE), also known as metallorganic CVD (MOCVD).

In this technique the required elements are transported as components of gaseous compounds such as metal alkyls and non-metal hydrides (e.g., trimethylgallium $(CH_3)_3Ga$, arsine AsH_3) to a suitable chamber where they flow over the surface of a heated substrate. These compounds break down and react so as to deposit the relevant semiconductor, with the remaining waste gases being removed from the chamber. MOVPE is used extensively for the production of compound semiconductor (e.g., III-V) thin films, such as the production of GaAs and AlGaAs layers using trimethyl gallium (TMG):

$$Ga(CH_3)_3 + AsH_3(g) \rightarrow GaAs(s) + 3CH_4(g);$$

$$xAl(CH_3)_3(g) + (1 - x)Ga(CH_3)3(g) + AsH_3(g)$$

$$\rightarrow Al_xGa_{1-x}As(s) + 3CH_4(g)$$

Valves in the gas lines allow gas switching, thereby enabling layered structures to be deposited. However, since rapid switching of a gas is more difficult than for an atomic beam, and because of residual gas adsorption on the walls of the feed pipework, the interfaces between adjacent layers grown by the gas-source epitaxy methods are not as abrupt as those attainable with solid-source MBE. For the same reason, solid-source MBE is capable of growing thinner layers, and the slow growth rates involved (of order 1 mm per hour) generally give rise to higher precision in layer thicknesses compared to MOVPE and MOMBE.

On the other hand, the high growth rate of MOVPE is better suited for commercial production, especially for semiconductor lasers where thick cladding or waveguiding layers must be grown in a reasonable timescale, in addition to the thin quantum well layers. MOVPE is also easier to scale up to larger systems, permitting simultaneous growth on multiple substrates. Gas source MBE, including MOMBE, generally operates with speeds and precisions which are intermediate between those of solid-source MBE and MOVPE. All the metallorganic epitaxy techniques have significant safety implications as the gases used are highly toxic.

The aim of molecular beam epitaxy is to produce a perfect single-crystal structure whose morphology and lattice spacing exactly match those of the substrate material. This requires the equilibrium lattice constant of all the different epitaxial layers to be closely matched to that of the substrate, otherwise dislocations will form. For some combinations of layers, a suitable substrate cannot be identified, in which case an initial buffer layer is deposited to serve as a 'virtual' substrate. In some material systems the deposited layers are sometimes annealed to produce an atomically smooth surface prior to further deposition of another material. The prevalence of each mode depends on the exact substrate temperature and deposition rate.

1D step propagation growth occurs at high temperatures and/or low growth rates. It occurs on vicinal, stepped surfaces when the surface diffusion of the growing species is pronounced. Deposited material diffuses to a step and is incorporated into the laterally growing terrace. This growth mode produces very flat and smooth interfaces.

2D nucleation and growth occurs at intermediate temperatures and/or growth rates and involves the initial nucleation of islands, which grow and coalesce to form a network of interconnected islands separated by holes. These holes are finally filled and a flat surface is formed. The process then repeats in a cyclical fashion.

3D rough growth generally occurs at low temperatures and high deposition rates. Islands nucleate on existing islands and extensive faceting occurs, producing very rough interfaces.

Because the dopants in both MBE and MOVPE systems exist as additional switchable sources, they may be added in known amounts and distributed over known depth concentration profiles such an abrupt delta-doped layers or extended and/or graded distributions. Both MBE andMOVPE are very successful in growing quantum well structures with a sensitive control of layer thickneses. However, it is much more difficult to achieve fine control of lateral dimensions in devices for the production of either quantum wires or quantum dots.

One technique for achieving lateral control is by the use of patterning. Patterning can be carried out before or after epitaxial growth (or both) using lithography or etching. In addition, under certain growth conditions, quantum wires or dotsmay form spontaneously during the epitaxial growth of strained materials.

A variant of many of the PVD processes described above are thermal spraying techniques in which a spray of molten or semi-molten solid particles generated by either an electrical thermal source (e.g., plasma spraying) or by chemical combustion (e.g., flame spraying or high-velocity oxygen fuel spraying) are deposited onto a substrate and undergo rapid solidification.

More generally, liquid phase chemical synthesis involves the reaction of a solution of precursor chemicals in either an aqueous or non-aqueous solvent; these precursors react and/or naturally self-assemble to form a solution supersaturated with the product. Thermodynamically this is an unstable situation and ultimately results in nucleation of the product either homogeneously (in solution) or heterogeneously.

Initial nuclei then grow into nanometre-sized particles or architectures according to both thermodynamic and kinetic factors. The nature, size and morphological shape of the precipitated structures can often be controlled by parameters such as temperature, pH, reactant concentration and time. A multicomponent product may require careful control of co-precipitation conditions in order to achieve a chemically homogeneous final product.

3

Developments in Nanotechnology

Nano-science and -technology are therefore both seen as (i) dealing with structures smaller than 100 nm in at least one dimension and (ii) as describing the deliberate manufacture and/or manipulation of individual nanostructures. NanoS&T requires a high degree of interdisciplinary and transdisciplinary cooperation and communication.

This is due to the fact that, at the nano level, the terminology of physics, chemistry and biology overlaps and blurs, and also to the fact that techniques from a single discipline can or must be supplemented by techniques and expertise from other disciplines. Nanotechnology is also an enabling technology which can impact on other fields leading to new developments and challenges. NanoS&T is not just about miniaturising things.

At the nanoscale, different laws of physics (quantum physics) become dominant, the properties of traditional materials change and the behaviour of surfaces starts to dominate the behaviour of bulk materials. Production methods for nanomaterials are often very different from those for the traditional materials. In addition, making materials atomically exact can change their properties radically. Basic structures of nanomaterials are:

— nanoparticles: ultrafine solid particles on a nanoscale including nanopowders and nanocrystals;

- nanotubes: hollow nanoscale particles including nanotubes, nanohorns and nanocapsules;
- nanostructured materials and coatings: materials made of structural elements (clusters, crystals, molecules) with dimensions in the nano-range and which may form films or be free standing; and
- nanocomposites: mixtures of components at least one of which has nanoscale dimensions.

Traditional bulk materials for manufacturing are generally uncontrolled and disordered at small scales. By controlling the nanostructure of materials, order can be introduced and novel mechanical, thermal, electrical, magnetic and other properties can be engineered. For instance, metal alloys are made of crystals whose size and shape is only crudely controlled. In comparison, a carbon nanotube (a tiny, hollow tube of carbon atoms) can be perfectly formed, is remarkably strong, and has useful electrical and thermal properties.

When particles become small enough to be nanoparticles, their mechanical properties change, and the way light and other electromagnetic radiation is affected by them changes (visible light wavelengths are on the order of a few hundred nanometers). Using nanoparticles in composite materials can enhance their strength, reduce weight, increase chemical and heat resistance and change their interaction with radiation.

Coatings made from nanoparticles can be unusually tough or slippery, or exhibit unusual properties, such as changing colour when a current is applied or cleaning themselves when it rains. Although hard to quantify, it has been estimated that nanostructured materials and processes can be expected to have a market impact of over \$ 1 trillion (\$ 10^{12}) by 2015.

The high level of uncertainty in the predicted figure comes from not knowing what materials will be developed and what their potential applications will be. In the short term, industry is seeking to produce low cost, reproducible manufactured materials with acceptable levels of associated risk (e.g. environmental effects, toxicity). Some existing applications include:

— clay nanoparticles in packaging materials, where reduced porosity leads to less gas entering (e.g. less gas such as oxygen that spoils foods);

— rolled graphite nanotubes used in coatings on car bumpers that better hold their shape in a crash;

— carbon nanotubes which are sources of field-emitted electrons and create enhanced phosphorescence e.g. in "jumbotron" lamps used at many athletic stadiums

— nanoparticles of zinc oxide in sunscreens, more efficient at absorbing UV than more traditional white titanium dioxide lotions and leaving the lotion smooth and transparent;

— textiles which are dirt and crease resistant due to nanocoatings;

— nanoparticles used as antiseptics, for abrasives and in paints;

— nanocoatings for spectacle glasses (making them scratchproof and crack resistant)

— nanocoatings for tiles to reduce slipping;

— electrochromic or self-cleaning nanofilm coatings on windows, which in sunshine breaks down dirt and helps the water falling on it to carry the dirt away;

— nanofilms with non-stick properties used as anti-graffiti coatings for walls;

— ceramic coatings for solar cells to improve scratch and erosion resistance;

— glues containing nanoparticles with variable optical properties are used in optoelectronics (e.g. for coupling fibres to other optical components); and

— conductive nanofilms used in casings for electronic devices, such as computers, to provide shielding against electromagnetic interference.

Applications under development, some of which are close to market, include:

— "smart" fabrics that can change their physical properties according to surrounding conditions, or even monitor vital signs;

— drug delivery mechanisms including antibacterial and antiviral nanoparticles;

— nanoceramics for more durable and better medical prosthetics;

— improved catalysts for fuel production; and

— nanoengineered membranes for energy efficient water purification.

Nanotechnology and Life on Earth

Nanotechnology can help life spread beyond Earth—a step without parallel since life spread beyond the seas. It can help mind emerge in machines—a step without parallel since mind emerged in primates. And it can let our minds renew and remake our bodies—a step without any parallel at all. These revolutions will bring dangers and opportunities too vast for the human imagination to grasp. Yet the principles of change that have applied to molecules, cells, beasts, minds, and machines should endure even in an age of biotechnology, nanomachines, and artificial minds. The same principles that have applied at sea, on land, and in the air should endure as we spread Earth's life toward the stars. Understanding the enduring principles of change will help us understand the potential for good and ill in the new technologies.

In crystal growth, each layer forms a template for the next. Uniform layers accumulate to form a solid block. In cells, strands of DNA or RNA can serve as templates too, aided by enzymes that act as molecular copying machines. But the subunits of nucleic acid strands can be arranged in many different sequences, and a template strand can separate from its copy.

Both strand and copy can then be copied again. Biochemist Sol Spiegelman has used a copying machine in test tube experiments. In a simple, lifeless environment, it duplicates RNA molecules. Picture a strand of RNA floating in a test tube together with copying machines and RNA subunits. The strand tumbles and writhes until it bumps into a copying machine in the right position to stick. Subunits bump around until one of the right kind meets the copying machine in the right position to match the template strand.

As matching subunits chance to fall into position, the machine seizes them and bonds them to the growing copy; though subunits bump randomly, the machine bonds selectively. Finally the machine, the template, and the copy separate. In the terminology of Oxford zoologist Richard Dawkins, things that give rise to copies of themselves are called replicators. In this environment, RNA molecules qualify: a single molecule soon becomes two, then four, eight, sixteen, thirty-two, and so forth, multiplying exponentially. Later, the replication rate levels off: the fixed stock of protein machines can churn out RNA copies only so fast, no matter how many template molecules vie for their services.

Later still, the raw materials for making RNA molecules become scarce and replication starves to a halt. The exploding population of molecules reaches a limit to growth and stops reproducing. The copying machines, however, often miscopy an RNA strand, inserting, deleting, or mismatching a subunit. The resulting mutated strand then differs in length or subunit sequence. Such changes are fairly random, and changes accumulate as miscopied molecules are again miscopied. As the molecules proliferate, they begin to grow different from their ancestors and from each other. This might seem a recipe for chaos. Biochemists have found that differing RNA molecules replicate at differing rates, depending on their lengths and subunit patterns.

Descendants of the swifter replicators naturally grow more common. Indeed, if one kind replicates just 10 percent more rapidly than its siblings, then after one hundred generations, each of the faster kind gives rise to 1,000 times as many descendants. Small differences in exponential growth pile up exponentially. When a test tube runs out of subunits, an experimenter can sample its RNA and "infect" a fresh tube. The process begins again and the molecules that dominated the first round of competition begin with a head start. More small changes appear, building over time into large changes.

Some molecules replicate faster, and their kind dominates the mix. When resources run out, the experimenter can sample the RNA and start again (and again, and again), holding conditions stable. This experiment reveals a natural process: no matter what RNA sequences the experimenter starts with, the seeming chaos of random errors and biased copying brings forth one kind of RNA molecule. Its typical version has a known, well-defined sequence of 220 subunits. It is the best RNA replicator in this environment, so it crowds out the others and stays.

Prolonged copying, miscopying, and competition always bring about the same result, no matter what the length or pattern of the RNA molecule that starts the process. Though no one could have predicted this winning pattern, anyone can see that change and competition will tend to bring forth a single winner. Little else could happen in so simple a system. If these replicators affected one another strongly, then the result could resemble a more complex ecology. As it is, they just compete for a resource. A variation on this example shows us something else: RNA molecules adapt differently to different environments. A molecular machine called a ribonuclease grabs RNA molecules having certain sequences of exposed subunits and cuts them in two. But RNA molecules, like proteins, fold in patterns that depend on their sequences, and by folding the right way they can protect their vulnerable spots.

Experimenters find that RNA molecules evolve to sacrifice swift replication for better protection when ribonuclease is around. Again, a best competitor emerges. Notice that biological terms have crept into this description: since the molecules replicate, the word "generation" seems right; molecules "descended" from a common "ancestor" are "relatives," and the words "growth," "reproduction," "mutation," and "competition" also seem right. Why is this? Because these molecules copy themselves with small variations, as do the genes of living organisms. When varying replicators have varying successes, the more successful tend to accumulate.

This process, wherever it occurs, is "evolution." In this test tube example we can see evolution stripped to its bare essentials, free of the emotional controversy surrounding the evolution of life. The RNA replicators and protein copying machines are well-defined collections of atoms obeying well-understood principles and evolving in repeatable laboratory conditions.

Biochemists can make RNA and protein from off-the-shelf chemicals, without help from life. Biochemists borrow these copying machines from a kind of virus that infects bacteria and uses RNA as its genetic material. These viruses survive by entering a bacterium, getting themselves copied using its resources, and then escaping to infect new bacteria. Miscopying of viral RNA produces mutant viruses, and viruses that replicate more successfully grow more common; this is evolution by natural selection, apparently called "natural" because it involves nonhuman parts of nature. But unlike the test tube RNA, viral RNA must do more than just replicate itself as a bare molecule, Successful viral RNA must also direct bacterial ribosomes to build protein devices that let it first escape from the old bacterium, then survive outside, and finally enter a new one. This additional information makes viral RNA molecules about 4,500 subunits long.

To replicate successfully, the DNA of large organisms must do even more, directing the construction of tens of thousands of different protein machines and the development of complex tissues and organs. This requires thousands of genes coded in millions to billions of DNA subunits. Nevertheless, the essential process of evolution by variation and selection remains the same in the test tube, in viruses, and far beyond. There are at least three ways to explain the structure of an evolved population of molecular replicators, whether test tube RNA, viral genes, or human genes.

The first kind of explanation is a blow-by-blow account of their histories: how specific mutations occurred and how they spread. This is impossible without recording all the molecular events, and such a record would in any event be immensely tedious. The second kind of explanation resorts to a somewhat misleading word: purpose. In detail, the molecules simply change haphazardly and replicate selectively. Yet stepping back from the process, one could describe the outcome by imagining that the surviving molecules have changed to "achieve the goal" of replication.

Why do RNA molecules that evolved under the threat of ribonuclease fold as they do? Because of a long and detailed history, of course, but the idea that "they want to avoid attack and survive to replicate" would predict the same result. The language of purpose makes useful shorthand, but the appearance of purpose need not result from the action of a mind. The RNA example shows this quite neatly. The third kind of explanation—in terms of evolution—says that order emerges through the variation and selection of replicators. A molecule folds in a particular way because it resembles ancestors that multiplied more successfully (by avoiding attack, etc.), and left descendants including itself.

As Richard Dawkins points out, the language of purpose (if used carefully) can be translated into the language of evolution. Evolution attributes patterns of success to the elimination of unsuccessful changes. It thus explains a positive as the result of a double negative—an explanation of a sort that seems slightly difficult to grasp. Worse, it explains something visible in terms of something invisible. Because only successful beasts have littered the landscape with the bones of their descendants, the malformed failures of the past haven't even left many fossils. The human mind tends to focus on the visible, seeking positive causes for positive results, an ordering force behind orderly results. Yet through reflection we can see that this great principle has changed our past and will shape our future: Evolution proceeds by the variation and selection of replicators.

Organisms

The history of life is the history of an arms race based on molecular machinery. Today, as this race approaches a new and swifter phase, we need to be sure we understand just how deeply rooted evolution is. In a time when the idea of biological evolution is often slighted in the schools and sometimes attacked, we should remember that the supporting evidence is as solid as rock and as common as cells. In pages of stone, the Earth itself has recorded the history of life. On lake bottoms and seabed, shells, bones, and silt have piled, layer on layer. Sometimes a shifting current or a geological upheaval has washed layers away; otherwise they have simply deepened. Early layers, buried deep, have been crushed, baked, soaked in mineral waters, and turned to stone. For centuries, geologists have studied rocks to read Earth's past. Long ago, they found seashells high in the crushed and crumpled rock of mountain ranges. By 1785 - seventy-four years before Darwin's detested book - James Hutton had concluded that seabed mud had been pressed to stone and raised skyward by forces not yet understood. What else could geologists think, unless nature itself had lied? They saw that fossil bones and shells differed from layer to layer. They saw that shells in layers here matched shells in layers there, though the layers might lie deep beneath the land between.

They named layers (A,B,C,D..., or Osagian, Meramecian, Lower Chesterian, Upper Chesterian...), and used characteristic fossils to trace rock layers. The churning of Earth's crust has nowhere left a complete sequence of layers exposed, yet geologists finding A,B,C,D,E in one place, C,D,E,F,G,H,I,J in another and J,K,L somewhere else could see that A preceded L. Petroleum geologists still use such fossils to date rock layers and to trace layers from one drill site to another.

Scientists came to the obvious conclusion. Just as sea species today live in broad areas, so did species in years gone by. Just as layer piles on top of layer today, so did they then. Similar shells in similar layers mark sediments laid down in the same age. Shells change from layer to layer because species changed from age to age. This is what geologists found written in shells and bones on pages of stone. The uppermost layers of rock contain bones of recent animals, deeper layers contain bones of animals now extinct. Still earlier layers show no trace of any modern species.

Below mammal bones lie dinosaur bones; in older layers lie amphibian bones, then shells and fish bones, and then no bones or shells at all. The oldest fossilbearing rocks bear the microscopic traces of single cells. Radioactive dating shows these oldest traces to be several billion years old. Cells more complex than bacteria date to little more than one billion years ago. The history of worms, fish, amphibians, reptiles, and mammals spans hundreds of millions of years. Human-like bones date back several million years.

The remains of civilisations date back several thousand. In three billion years, life evolved from single cells able to soak up chemicals to collections of cells embodying minds able to soak up ideas. Within the last century, technology has evolved from the steam locomotive and electric light to the spaceship and the electronic computer - and computers are already being taught to read and write. With mind and technology, the rate of evolution has jumped a millionfold or more. The book of stone records the forms of long-dead organisms, yet living cells also carry records, genetic texts only now being read.

As with the ideas of geology, the essential ideas of evolution were known before Darwin had set pen to paper. In lamp-lit temples and monasteries, generations of scribes copied and recopied manuscripts. Sometimes they miscopied words and sentences - whether by accident, by perversity, or by order of the local ruler - and as the manuscripts replicated, aided by these human copying machines, errors accumulated.

The worst errors might be caught and removed, and famous passages might survive unchanged, but differences grew. Ancient books seldom exist in their original versions. The oldest copies are often centuries younger than the lost originals. Nonetheless, from differing copies with differing errors, scholars can reconstruct versions closer to the original. They compare texts. They can trace lines of descent from common ancestors because unique patterns of errors betray copying from a common source. Where all surviving copies agree, scholars can assume that the original copy held the same words. Where survivors differ, scholars study copies that descended separately from a distant ancestor, because areas of agreement then indicate a common origin in the ancestral version. Genes resemble manuscripts written in a four-letter alphabet. Much as a message can take many forms in ordinary language, so different genetic wording can direct the construction of identical protein molecules. Moreover, protein molecules with different design details can serve identical functions.

A collection of genes in a cell is like a whole book, and genes - like old manuscripts - have been copied and recopied by inaccurate scribes. Like scholars studying ancient texts, biologists generally work with modern copies of their material. They compare organisms with similar appearances and find that they give similar answers to the essay questions in their genes and proteins. The more two organisms differ (lions and lizards, humans and sunflowers), the more these answers differ, even among molecular machines serving identical functions.

More telling still, similar animals make the same mistakes - all primates, for example, lack enzymes for making vitamin C, an omission shared by only two other known mammals, the guinea pig and the fruit bat. This suggests that we primates have copied our genetic answers from a shared source, long ago. The same principle that shows the lines of descent of ancient texts thus also reveals the lines of descent of modern life. Indeed, it indicates that all known life shares a common ancestor.

The first replicators on Earth evolved abilities beyond those possible to RNA molecules replicating in test tubes. By the time they reached the bacterial stage, they had developed the "modern" system of using DNA, RNA, and ribosomes to construct protein. Mutations then changed not only the replicating DNA itself, but protein machines and the living structures they build and shape. Teams of genes shaped ever more elaborate cells, then guided the cellular cooperation that formed complex organisms. Variation and selection favoured teams of genes that shaped beasts with protective skins and hungry mouths, animated by nerve and muscle, guided by eye and brain. As Richard Dawkins puts it, genes built ever more elaborate survival machines to aid their own replication. When dog genes replicate, they often shuffle with those of other dogs that have been selected by people, who then select which puppies to keep and breed.

Over the millennia, people have molded wolf-like beasts into greyhounds, toy poodles, dachshunds, and Saint Bernards. By selecting which genes survive, people have reshaped dogs in both body and temperament. Human desires have defined success for dog genes; other pressures have defined success for wolf genes. Mutation and selection of genes has, through long ages, filled the world with grass and trees, with insects, fish, and people. More recently, other things have appeared and multiplied - tools, houses, aircraft, and computers. And like the lifeless RNA molecules, this hardware has evolved.

Technological Variations

As the stone of Earth records the emergence of ever more complex and capable forms of life, so the relics and writings of humanity record the emergence of ever more complex and capable forms of hardware. Our oldest surviving hardware is itself stone, buried with the fossils of our ancestors; our newest hardware orbits overhead. Consider for a moment the hybrid ancestry of the space shuttle. On its aircraft side, it descends from the aluminum jets of the sixties, which themselves sprang from a line stretching back through the aluminum prop planes of World War II, to the wood-and-cloth biplanes of World War I, to the motorised gliders of the Wright brothers, to toy gliders and kites. On its rocket side, the shuttle traces back to Moon rockets, to military missiles, to last century's artillery rockets ("and the rocket's red glare..."), and finally to fireworks and toys. This aircraft/rocket hybrid flies, and by varying components and designs, aerospace engineers will evolve still better ones. Generation and testing of alternatives is synonymous with variation and selection.

Sometimes various alternatives already exist. In "One Highly Evolved Toolbox," in The Next Whole Earth Catalog, J. Baldwin writes: "Our portable shop has been evolving for about twenty years now. There's nothing really very special about it except that a continuing process of removing obsolete or inadequate tools and replacing them with more suitable ones has resulted in a collection that has become a thing-making system rather than a pile of hardware." Baldwin uses the term "evolving" accurately.

Invention and manufacture have for millennia generated variations in tool designs, and Baldwin has winnowed the current crop by competitive selection, keeping those that work best with his other tools to serve his needs. Through years of variation and selection, his system evolved - a process he highly recommends. Indeed, he urges that one never try to plan out the purchase of a complete set of tools. Instead, he urges buying the tools one often borrows, tools selected not by theory but by experience.

Technological variations are often deliberate, in the sense that engineers are paid to invent and test. Still, some novelties are sheer accident, like the discovery of a crude form of Teflon in a cylinder supposedly full of tetrafluoroethylene gas: with its valve open, it remained heavy; when it was sawed open, it revealed a strange, waxy solid. Other novelties have come from systematic blundering. Edison tried carbonising everything from paper to bamboo to spider webs when he was seeking a good light-bulb filament. Charles Good year messed around in a kitchen for years, trying to convert gummy natural rubber into a durable substance, until at last he chanced to drop sulfurised rubber on a hot stove, performing the first crude vulcanisation. In engineering, enlightened trial and error, not the planning of flawless intellects, has brought most advances; this is why engineers build prototypes.

Peters and Waterman in their book In Search of Excellence show that the same holds true of advances in corporate products and policies. This is why excellent companies create "an environment and a set of attitudes that encourage experimentation," and why they evolve "in a very Darwinian way." Factories bring order through variation and selection. Crude quality-control systems test and discard faulty parts before assembling products, and sophisticated quality-control systems use statistical methods to track defects to their sources, helping engineers change the manufacturing process to minimise defects.

Japanese engineers, building on W. Edwards Deming's work in statistical quality control, have made such variation and selection of industrial processes a pillar of their country's economic success. Assembler-based systems will likewise need to measure results to eliminate flaws. Quality control is a sort of evolution, aiming not at change but at eliminating harmful variations. But just as Darwinian evolution can preserve and spread favourable mutations, so good quality control systems can help managers and workers to preserve and spread more effective processes, whether they appear by accident or by design. All this tinkering by engineers and manufacturers prepares products for their ultimate test. Out in the market, endless varieties of wrench, car, sock, and computer compete for the favour of buyers.

When informed buyers are free to choose, products that do too little or cost too much eventually fail to be reproduced. As in nature, competitive testing makes yesterday's best competitor into tomorrow's fossil. "Ecology" and "economy" share more than linguistic roots. Both in the marketplace and on real and imaginary battlefields, global competition drives organisations to invent, buy, beg, and steal ever more capable technologies. Some organisations compete chiefly to serve people with superior goods, others compete chiefly to intimidate them with superior weapons. The pressures of evolution drive both. The global technology race has been accelerating for billions of years.

The earthworm's blindness could not block the development of sharp-eyed birds. The bird's small brain and clumsy wings could not block the development of human hands, minds, and shotguns. Likewise, local prohibitions cannot block advances in military and commercial technology. It seems that we must guide the technology race or die, yet the force of technological evolution makes a mockery of antitechnology movements: democratic movements for local restraint can only restrain the world's democracies, not the world as a whole.

Evolution in design

It might seem that design offers an alternative to evolution, but design involves evolution in two distinct ways. First, design practice itself evolves. Not only do engineers accumulate designs that work, they accumulate design methods that work. These range from handbook standards for choosing pipes to management systems for organising research and development. And as Alfred North Whitehead stated, "The greatest invention of the nineteenth century was the invention of the method of invention." Second, design itself proceeds by variation and selection, Engineers often use mathematical laws evolved to describe heat flow and elasticity to test simulated designs before building them. They thus evolve plans through a cycle of design, calculation, criticism, and redesign, avoiding the expense of cutting metal.

The creation of designs thus proceeds through a nonmaterial form of evolution. Hooke's law, for example, describes how metal bends and stretches: deformation is proportional to the applied stress: twice the pull, twice the stretch. Though only roughly correct, it remains fairly accurate until the metal's springiness finally yields to stress.

Engineers can use a form of Hooke's law to design a bar of metal that can support a load without bending too far—and then make it just a bit thicker to allow for inaccuracies in the law and in their design calculations. They can also use a form of Hooke's law to describe the bending and twisting of aircraft wings, tennis rackets, and automobile frames. But simple mathematical equations don't wrap smoothly around such convoluted structures. Engineers have to fit the equations to simpler shapes (to pieces of the design), and then assemble these partial solutions to describe the flexing of the whole. It is a method that typically requires immense calculations, and without computers it would be impractical.

With them, it has grown common. Such simulations extend an ancient trend. We have always imagined consequences, in hope and fear, when we have needed to select a course of action. Simpler mental models (whether inborn or learned) undoubtedly guide animals as well. When based on accurate mental models, thought experiments can replace more costly physical experiments—a development evolution has favoured.

Engineering simulations simply extend this ability to imagine consequences, to make our mistakes in thought rather than deed. A craftsman in a job shop can keep tools in plain sight; working with them every day makes them familiar to his eyes, hands, and mind. He gets to know their abilities naturally, and can put this knowledge to immediate creative use. But people - like us - who have to understand the future face a greater challenge, because the future's tools exist now only as ideas and as possibilities implicit in natural law. These tools neither hang on the wall nor impress themselves on the mind through sight and sound and touch - nor will they, until they exist as hardware. In the coming years of preparation only study, imagination, and thought can make their abilities real to the mind.

New Replicators

History shows us that hardware evolves. Test tube RNA, viruses, and dogs all show how evolution proceeds by the modification and testing of replicators. But hardware (today) cannot reproduce itself - so where are the replicators behind the evolution of technology? What are the machine genes? Of course, we need not actually identify replicators in order to recognise evolution.

Darwin described evolution before Mendel discovered genes, and geneticists learned much about heredity before Watson and Crick discovered the structure of DNA. Darwin needed no knowledge of molecular genetics to see that organisms varied and that some left more descendants. A replicator is a pattern that can get copies of itself made. It may need help; without protein machines to copy it, DNA could not replicate. But by this standard, some machines are replicators! Companies often make machines that fall into the hands of a competitor; the competitor then learns their secrets and builds copies. Just as genes "use" protein machines to replicate, so such machines "use" human minds and hands to replicate.

With nanocomputers directing assemblers and disassemblers, the replication of hardware could even be automated. The human mind, though, is a far subtler engine of imitation than any mere protein machine or assembler. Voice, writing, and drawing can transmit designs from mind to mind before they take form as hardware. The ideas behind methods of design are subtler yet: more abstract than hardware, they replicate and function exclusively in the world of minds and symbol systems. Where genes have evolved over generations and eons, mental replicators now evolve over days and decades. Like genes, ideas split, combine, and take multiple forms. Science cannot yet describe the neural patterns that embody ideas in brains, but anyone can see that ideas mutate, replicate, and compete. Ideas evolve.

Richard Dawkins calls bits of replicating mental patterns "memes". He says "examples of memes are tunes, ideas, catch-phrases, clothes fashions, ways of making pots or of building arches. Just as genes propagate themselves in the gene pool by leaping from body to body via sperms or eggs, so memes propagate themselves in the meme pool by leaping from brain to brain via a process which, in the broad sense, can be called imitation."Memes replicate because people both learn and teach. They vary because people create the new and misunderstand the old. They are selected (in part) because people don't believe or repeat everything they hear.

As test tube RNA molecules compete for scarce copying machines and subunits, so memes must compete for a scarce resource - human attention and effort. Since memes shape behaviour, their success or failure is a deadly serious matter. Since ancient times, mental models and patterns of behaviour have passed from parent to child. Meme patterns that aid survival and reproduction have tended to spread. Year by year, people varied their actions with varying results. Year by year, some died while others found new tricks of survival and passed them on. Genes built brains skilled at imitation because the patterns imitated were, on the whole, of value - their bearers, after all, had survived to spread them.

Memes themselves, though, face their own matters of "life" and "death": as replicators, they evolve solely to survive and spread. Like viruses, they can replicate without aiding their host's survival or well-being. Indeed, the meme for martyrdom-in-a-cause can spread itself through the very act of killing its host. Genes, like memes, survive by many strategies. Some duck genes have spread themselves by encouraging ducks to pair off to care for their gene-bearing eggs and young. Some duck genes have spread themselves (when in male ducks) by encouraging rape, and some (when in female ducks) by encouraging the planting of eggs in other ducks' nests. Still other genes found in ducks are virus genes, able to spread without making more ducks.

Protecting eggs helps the duck species survive; rape helps one set of duck genes at the expense of others; infection helps viral genes at the expense of duck genes in general. As Richard Dawkins points out, genes "care" only about their own replication: they appear selfish. But selfish motives can encourage cooperation. People seeking money and recognition for themselves cooperate to build corporations that serve other people's wants.

Selfish genes cooperate to build organisms that themselves often cooperate. Even so, to imagine that genes automatically serve some greater good is to mistake a common effect for an underlying cause. To ignore the selfishness of replicators is to be lulled by a dangerous illusion. Some genes in cells are out-and-out parasites. Like herpes genes inserted in human chromosomes, they exploit cells and harm their hosts. Yet if genes can be parasites, why not memes as well?

In The Extended Phenotype, Richard Dawkins describes a worm that parasitises bees and completes its life cycle in water. It gets from bee to water by making the host bee dive to its death. Similarly, ant brainworms must enter a sheep to complete their life cycle. To accomplish this, they burrow into the host ant's brain, somehow causing changes that make the ant "want" to climb to the top of a grass stem and wait, eventually to be eaten by a sheep.

As worms enter other organisms and use them to survive and replicate, so do memes. Indeed, the absence of memes exploiting people for their own selfish ends would be amazing, a sign of some powerful - indeed, nearly perfect - mental immune system. But parasitic memes clearly do exist. Just as viruses evolve to stimulate cells to make viruses, so rumours evolve to sound plausible and juicy, stimulating repetition. Ask not whether a rumour is true, ask instead how it spreads. Experience shows that ideas evolved to be successful replicators need have little to do with the truth. At best, chain letters, spurious rumours, fashionable lunacies, and other mental parasites harm people by wasting their time. At worst, they implant deadly misconceptions. These meme systems exploit human ignorance and vulnerability. Spreading them is like having a cold and sneezing on a friend. Though some memes act much like viruses, infectiousness isn't necessarily bad (think of an infectious grin, or infectious good nature). If a package of ideas has merit, then its infectiousness simply increases its merit - and indeed, the best ethical teachings also teach us to teach ethics.

Spreading useful meme systems is like offering useful seeds to a friend with a garden. Parasites have forced organisms to evolve immune systems, such as the enzymes that bacteria use to cut up invading viruses, or the roving white blood cells our bodies use to destroy bacteria. Parasitic memes have forced minds down a similar path, evolving meme systems that serve as mental immune systems.

The oldest and simplest mental immune system simply commands "believe the old, reject the new." Something like this system generally kept tribes from abandoning old, tested ways in favour of wild new notions - such as the notion that obeying alleged ghostly orders to destroy all the tribe's cattle and grain would somehow bring forth a miraculous abundance of food and armies of ancestors to drive out foreigners. Your body's immune system follows a similar rule: it generally accepts all the cell types present in early life and rejects new types such as potential cancer cells and invading bacteria, as foreign and dangerous. This simple reject-the-new system once worked well, yet in this era of organ transplantation it can kill.

Similarly, in an era when science and technology regularly present facts that are both new and trustworthy, a rigid mental immune system becomes a dangerous handicap. For all its shortcomings, though, the reject-the-new principle is simple and offers real advantages. Tradition holds much that is tried and true.

Change is risky: just as most mutations are bad, so most new ideas are wrong. Even reason can be dangerous: if a tradition links sound practices to a fear of ghosts, then overconfident rational thought may throw out the good with the bogus. Unfortunately, traditions evolved to be good may have less appeal than ideas evolved to sound good - when first questioned, the soundest tradition may be displaced by worse ideas that better appeal to the rational mind. Yet memes that seal the mind against new ideas protect themselves in a suspiciously self-serving way. While protecting valuable traditions from clumsy editing, they may also shield parasitic claptrap from the test of truth.

In times of swift change they can make minds dangerously rigid. Much of the history of philosophy and science may be seen as a search for better mental immune systems, for better ways to reject the false, the worthless, and the damaging. The best systems respect tradition, yet encourage experiment. They suggest standards for judging memes, helping the mind distinguish between parasites and tools. The principles of evolution provide a way to view change, whether in molecules, organisms, technologies, minds, or cultures.

The deep-rooted principles of evolutionary change will shape the development of nanotechnology, even as the distinction between hardware and life begins to blur. These principles show much about what we can and cannot hope to achieve, and they can help us focus our efforts to shape the future. They also tell us much about what we can and cannot foresee, because they guide the evolution not only of hardware, but of knowledge itself.

Political and economic trends are notoriously fickle, and sheer chance rolls dice across continents. Even the comparatively steady advance of technology often eludes prediction. Prognosticators often guess at the times and costs required to harness new technologies. When they reach beyond outlining possibilities and attempt accurate predictions, they generally fail. Though the space shuttle was clearly possible, predictions of its cost and initial launch date were wrong by several years and billions of dollars.

Engineers cannot accurately predict when a technology will be developed, because development always involves uncertainties. But we have to try to predict and guide development. Though one technology can sometimes block the dangers of another, competing technologies often go in the same direction. Biochemists developed techniques for making DNA without programmable nanomachines, using shortcuts based on specific chemical tricks.

Winning technologies often succeed because of unobvious tricks and details. In the mid-1950s physicists could see that basic semiconductor principles made microcircuits physically possible, but foreseeing how they would be made - foreseeing the details of mask-making, resists, oxide growth, ion implantation, etching, and so forth, in all their complexity - would have been impossible. The nuances of detail and competitive advantage that select winning technologies make the technology race complex and its path unpredictable.

In a race toward the limits set by natural law, the finish line is predictable even if the path and the pace of the runners are not. Not human whims but the unchanging laws of nature draw the line between what is physically possible and what is not - no political act, no social movement can change the law of gravity one whit. So however futuristic they may seem, sound projections of technological possibilities are quite distinct from predictions.

They rest on timeless laws of nature, not on the vagaries of events. It is unfortunate that this insight remains rare. Without it, we stumble in a daze across the landscape of the possible, confusing mountains with mirages and discounting both. We look ahead with minds and cultures rooted in the ideas of more sluggish times, when both science and technological competition lacked their present strength and speed. We have only recently begun to evolve a tradition of technological foresight.

Science vs. Technology

The evolution of technology development is often confused with science in the public mind. This confusion muddles our efforts at foresight. Though engineers often tread uncertain ground, they are not doomed to do so, as scientists are. They can escape the inherent risks of proposing precise, universal scientific theories. Engineers need only show that under particular conditions particular objects will perform well enough.

A designer need know neither the exact stress in a suspension bridge cable nor the exact stress that will break it; the cable will support the bridge so long as the first remains below the second, whatever they may be. Though measurements cannot prove precise equality, they can prove inequality. Engineering results can thus be solid in a way that precise scientific theories cannot. Engineering results can even survive disproof of the scientific theories supporting them, when the new theory gives similar results. The case for assemblers, for example, will survive any possible refinements in our theory of quantum mechanics and molecular bonds. Predicting the content of new scientific knowledge is logically impossible because it makes no sense to claim to know already the facts you will learn in the future. Predicting the details of future technology, on the other hand, is merely difficult. Science aims at knowing, but engineering aims at doing; this lets engineers speak of future achievements without paradox. They can evolve their hardware in the world of mind and computation, before cutting metal or even filling in all the details of a design.

Scientists commonly recognise this difference between scientific foresight and technological foresight: they readily make technological predictions about science. Scientists could and did predict the quality of Voyager's pictures of Saturn's rings, for example, though not their surprising content. Indeed, they predicted the pictures' quality while the cameras were as yet mere ideas and drawings. Their calculations used well-tested principles of optics, involving no new science. Because science aims to understand how everything works, scientific training can be a great aid in understanding specific pieces of hardware. Still, it does not automatically bring engineering expertise; designing an airliner requires much more than a knowledge of the sciences of metallurgy and aerodynamics.

Scientists are encouraged by their colleagues and their training to focus on ideas that can be tested with available apparatus. The resulting short-term focus often serves science well: it keeps scientists from wandering off into foggy worlds of untested fantasy, and swift testing makes for an efficient mental immune system. Regrettably, though, this cultural bias toward short-term testing may make scientists less interested in long-term advances in technology. The impossibility of genuine foresight regarding science leads many scientists to regard all statements about future developments as "speculative" - a term that makes perfect sense when applied to the future of science, but little sense when applied to well-grounded projections in technology. But most engineers share similar leanings toward the short term.

They too are encouraged by their training, colleagues, and employers to focus on just one kind of problem: the design of systems that can be made with present technology or with technology just around the corner. Even long-term engineering projects like the space shuttle must have a technology cutoff date after which no new developments can become part of the basic design of the system. In brief, scientists refuse to predict future scientific knowledge, and seldom discuss future engineering developments. Engineers do project future developments, but seldom discuss any not based on present abilities. Yet this leaves a crucial gap: what of engineering developments firmly based on present science but awaiting future abilities?

This gap leaves a fruitful area for study. Imagine a line of development which involves using existing tools to build new tools, then using those tools to build novel hardware. Each set of tools may rest on established principles, yet the whole development sequence may take many years, as each step brings a host of specific problems to iron out. Scientists planning their next experiment and engineers designing their next device may well ignore all but the first step. Still, the end result may be foreseeable, lying well within the bounds of the possible shown by established science.

Recent history illustrates this pattern. Few engineers considered building space stations before rockets reached orbit, but the principles were clear enough, and space systems engineering is now a thriving field. Similarly, few mathematicians and engineers studied the possibilities of computation until computers were built, though many did afterward. So it is not too surprising that few scientists and engineers have yet examined the future of nanotechnology, however important it may become.

4

Nanotechnology and Industrial Development

On Earth, a person can walk or sail only to a certain distance, but no mysterious edge or barrier suddenly blocks travel. The Earth is simply round. The speed limit in space no more implies a "light barrier" than the distance limit on Earth implies a wall. Space itself - the vacuum that holds all energy and matter - has properties.

One of these is its geometry, which can be described by regarding time as a special dimension. This geometry makes the speed of light recede before an accelerating spaceship much as the horizon recedes before a moving sea ship: the speed of light, like the horizon, is always equally remote in all directions. But the analogy dies here - this similarity has nothing to do with the curvature of space. It is enough to remember that the limiting speed is nothing so crude or so breakable as a "light barrier." Objects can always go faster, just no faster than light. People have long dreamed of gravity control. In the 1962 edition of Profiles of the Future, Arthur C. In the general theory of relativity, Einstein described gravity as curvature in the space-time structure of the vacuum.

The mathematics describing this is elegant and precise, and it makes predictions that have passed every test yet contrived. Gravity is neither more nor less implacable than other forces. No one can make a boulder lose its gravity, but neither can anyone make an electron lose its electric charge or a current its magnetic field. We control electric and magnetic fields by moving the particles that create them; we can control gravitational fields similarly, by moving ordinary masses. It seems that we cannot learn the secret of gravity control because we already have it.

A child with a small magnet can lift a nail, using a magnetic field to overwhelm Earth's gravitational pull. But unfortunately for eager gravitational engineers, using gravity to lift a nail requires a tremendous mass. Engineers stir up electromagnetic waves by shaking electric charges back and forth in an antenna; one can stir up gravity waves by shaking a rock in the air. But again, the gravitational effect is weak. Though a onekilowatt radio station is nothing extraordinary, all the shaking and spinning of all the masses in the solar system put together fails to radiate as much as a kilowatt of gravity waves.

We understand gravity well enough; it simply isn't much use in building machines much lighter than the Moon. But devices using large masses do work. A hydroelectric dam is part of a gravity machine (the other part being the Earth) that extracts energy from falling mass. Machines using black holes will be able to extract energy from falling mass with over fifty percent efficiency, based on $E = mc^2$. Lowering a single bucket of water into a black hole would yield as much energy as pouring several trillion buckets of water through the generators of a kilometer-high dam.

Because the laws of gravity describe how the vacuum curves, they also apply to science-fiction style "space warps." It seems that tunnels from one point in space to another would be unstable, even if they could be created in the first place. This prevents future spaceships from reaching distant points faster than light by taking a shortcut around the intervening space, and this limit to travel in turn sets a limit to growth. Einstein's laws seem to give an accurate description of the overall geometry of the vacuum. If so, then the limits that result seem inescapable: you can get rid of almost anything, but not the vacuum itself.

Other laws and limits seem inescapable for similar reasons. In fact, physicists have increasingly come to regard all of physical law in terms of the structure of the vacuum. Gravity waves are a certain kind of ripple in the vacuum; black holes are a certain kind of kink. Likewise, radio waves are another kind of ripple in the vacuum, and elementary particles are other, very different kinds of kinks. In this view, there is only one substance in the universe - the vacuum—but one that takes on a variety of forms, including those patterns of particles that we call "solid matter." This view suggests the inescapable quality of natural law. If a single substance fills the universe, is the universe, then its properties limit all that we can do.

The strangeness of modern physics, however, leads many people to distrust it. The revolutions that brought quantum mechanics and relativity gave rise to talk of "the uncertainty principle," "the wave nature of matter," "matter being energy," and "curved space-time." An air of paradox surrounds these ideas and thus physics itself. It is understandable that new technologies should seem odd to us, but why should the ancient and immutable laws of nature turn out to be bizarre and shocking?

Our brains and languages evolved to deal with things vastly larger than atoms, moving at a tiny fraction of the speed of light. They do a tolerable job of this, though it took people centuries to learn to describe the motion of a falling rock. But we have now stretched our knowledge far beyond the ancient world of the senses. We have found things that seem bizarre - and some that are simply beyond our ability to visualise. But "bizarre" does not mean mysterious and unpredictable.

Mathematics and experiments still work, letting scientists vary and select theories, evolving them to fit even a peculiar reality. Human minds have proved remarkably flexible, but it is no great surprise to find that we cannot always visualise the invisible. Part of the reason that physics seems so strange is that people crave oddities, and tend to spread memes that describe things as odd. Some people favor ideas that coat the world in layers and fill it with grade-B mysteries. Natural'y, they favor and spread memes that make matter seem immaterial and quantum mechanics seem like a branch of psychology. Relativity, it is said, reveals that matter is really energy. This encourages a smiling vagueness about the mystery of the universe. It might be clearer to say that relativity reveals energy to be a form of matter: in all its forms, energy has mass. Indeed, lightsails work on this principle, through the impact of mass on a surface.

Light even comes packaged in particles. Consider also the Heisenberg uncertainty principle, and the related fact that "the observer always affects the observed." The uncertainty principle is intrinsic to the mathematics describing ordinary matter (giving atoms their very size), but the associated "effect of the observer" has been presented in some popular books as a magical influence of consciousness on the world. In fact, the core idea is more prosaic. Imagine looking at a dust mote in a light beam: When you observe the reflected light, you certainly affect it—your eye absorbs it. Likewise, the light (with its mass) affects the dust mote: it bounces off the mote, exerting a force.

The result is not an effect of your mind on the dust, but of the light on the dust. Though quantum measurement has peculiarities far more subtle than this, none involves the mind reaching out to change reality. Finally, consider the "twin paradox." Relativity predicts that, if one of a pair of twins flies to another star and back at near the speed of light, the travelling twin will be younger than the stay-at-home twin.

Indeed, measurements with accurate clocks demonstrate the time-slowing effect of rapids motion. But this is not a "paradox"; it is a simple fact of nature. In 1894 the eminent physicist Albert A. Michelson stated: "The more important fundamental laws and facts of physical science have all been discovered, and these are now so firmly established that the possibility of their ever being supplanted in consequence of new discoveries is exceedingly remote... Our future discoveries must be looked for in the sixth place of decimals." But in 1895, Roentgen discovered X rays.

In 1896, Becquerel discovered radioactivity. In 1897, Thomson discovered the electron. In 1905, Einstein formulated the special theory of relativity (and thus explained Michelson's own 1887 observations regarding the speed of light). In 1905, Einstein also presented the photon theory of light. In 1911, Rutherford discovered the atomic nucleus. In 1915, Einstein formulated the general theory of relativity. In 1924-30, de Broglie, Heisenberg, Bohr, Pauli, and Dirac developed the foundations of quantum mechanics. In 1929, Hubble announced evidence for the expansion of the universe.

In 1931, Michelson died. Michelson had made a memorable mistake. People still point to his statement and what followed to support the view that we shouldn't claim any firm understanding of natural law, or of the limits to the possible. The great revolution in physics has led some people to conclude that science will hold endless important surprises - even surprises important to engineers. The content of quantum mechanics was a surprise, yet before it appeared, physics was obviously and grossly incomplete. Before quantum mechanics, you could have walked up to any scientist, grinned maliciously, rapped on his desk and asked, "What holds this thing together? Why is it brown and solid, while air is transparent and gaseous?" Your victim might have said something vague about atoms and their arrangements, but when you pressed for an explanation, you would at best have gotten an answer like "Who knows? Physics can't explain matter yet!" Hindsight is all too easy, yet in a world made of matter, inhabited by material people using material tools, this ignorance of the nature of matter was a gap in human knowledge that Michelson should perhaps have noted. It was a gap not in "the sixth place of decimals" but in the first.

It is also worth observing the extent to which Michelson was right. The laws of which he spoke included Newton's laws of gravity and motion, and Maxwell's laws of electromagnetism. And indeed, under conditions common in engineering, these laws have been modified only "in the sixth place of decimals." Einstein's laws of gravity and motion match Newton's laws closely except under extreme conditions of gravitation and speed; the quantum electrodynamic laws of Feynman, Schwinger, and Tomonaga match Maxwell's laws closely except under extreme conditions of size and energy.

The revolutions of relativity and quantum mechanics changed our knowledge about matter and energy, but matter and energy themselves remained unchanged - they are real and care nothing for our theories. Physicists now use a single set of laws to describe how atomic nuclei and electrons interact in atoms, molecules, molecular machines, living things, planets, and stars.

These laws are not yet completely general; the quest for a unified theory of all physics continues. But as physicist Stephen W. Hawking states, "at the moment we have a number of partial laws which govern the behaviour of the universe under all but the most extreme conditions." And by an engineer's standards, those conditions are extraordinarily extreme. Physicists regularly announce new particles observed in the debris from extremely energetic particle collisions, but you cannot buy one of these new particles in a box.

And this is important to recognise, because if a particle cannot be kept, then it cannot serve as a component of a stable machine. Boxes and their contents are made of electrons and nuclei. Nuclei, in turn, are made of protons and neutrons. Hydrogen atoms have single protons as their nuclei; lead atoms have eighty-two protons and over a hundred neutrons.

Isolated neutrons disintegrate in a few minutes. Few other stable particles are known: photons - the particles of light - are useful and can be trapped for a time; neutrinos are almost undetectable and cannot even be trapped. These particles (save the photon) have matching antiparticles. All other known particles disintegrate in a few millionths of a second or less. Thus, the only known building blocks for hardware are electrons and nuclei; these building blocks ordinarily combine to form atoms and molecules. Yet despite the power of modern physics, our knowledge still has obvious holes.

The shaky state of elementary particle theory leaves some limits uncertain. We may find new stable, "boxable" particles, such as magnetic monopoles or free quarks; if so, they will doubtless have uses. We may even find a new long-range field or form of radiation, though this seems increasingly unlikely. Finally, some new way of smashing particles together may improve our ability to convert known particles into other known particles. But in general, complex hardware will require complex, stable patterns of particles.

Outside the environment of a collapsed star, this means patterns of atoms, which are well described by relativistic quantum mechanics. The frontiers of physics have moved on. On a theoretical level, physicists seek a unified description of the interactions of all possible particles, even the most short-lived particles. On an experimental level, they study the patterns of subatomic debris created by highenergy collisions in particle accelerators. So long as no new, stable, and useful particle comes out of such a collision - or turns up as the residue of past cosmic upheavals - atoms will remain the sole building blocks of stable hardware. And engineering will remain a game that is played with already known pieces according to already known rules.

New particles would add pieces, not eliminate rules. Is molecular machinery really the end of the road where miniaturisation is concerned? The idea that molecular machinery might be a step toward a smaller "nuclear machinery" seems natural enough. One young man heard of molecular technology and its ability to manipulate atoms, and immediately concluded that molecular technology could do almost anything, even turn falling nuclear bombs into harmless lead bricks at a distance. Molecular technology can do no such thing.

Turning plutonium into lead (whether at a distance or not) lies beyond molecular technology for the same reason that turning lead into gold lay beyond an alchemist's chemistry. Molecular forces have little effect on the atomic nucleus. The nucleus holds over 99.9 percent of an atom's mass and occupies about 1/1,000,000,000,000,000 of its volume. Compared to the nucleus, the rest of an atom (an electron cloud) is less than airy fluff. Trying to change a nucleus by poking at it with a molecule is even more futile than trying to flatten a steel ball bearing by waving a ball of cotton candy at it.

Molecular technology can sort and rearrange atoms, but it cannot reach into a nucleus to change an atom's type. Nanomachines may be no help in building nuclear-scale machines, but could such machines even exist? Apparently not, at least under any conditions we can create in a laboratory. Machines must have a number of parts in close contact, but close-packed nuclei repel each other with ferocity. When splitting nuclei blasted Hiroshima, most of the energy was released by the violent electrostatic repulsion of the freshly split halves. The well-known difficulty of nuclear fusion stems from this same problem of nuclear repulsion.

In addition to splitting or fusing, nuclei can be made to emit or absorb various types of radiation. In one technique, they are made to gyrate in ways that yield useful information, letting doctors make medical images based on nuclear magnetic resonance. But all these phenomena rely only on the properties of well-separated nuclei. Isolated nuclei are too simple to act as machines or electronic circuits. Nuclei can be forced close together, but only under the immense pressures found inside a collapsed star. Doing engineering there would present substantial difficulties, even if a collapsed star were har.dy.

This returns us to the basic question, What can we accomplish by properly arranging atoms? Some limits already seem clear. The strongest materials possible will have roughly ten times the strength of today's strongest steel wire. It seems that the vibrations of heat will, at ordinary pressures, tear apart the most refractory solids at temperatures around four thousand degrees Celsius (roughly fifteen hundred degrees cooler than the Sun's surface).

These brute properties of matter - strength and heat resistance-cannot be greatly improved through complex, clever arrangements of atoms. The best arrangements seem likely to be fairly simple and regular. Other fairly simple goals include transmitting heat, insulating against heat, transmitting electricity, insulating against electricity, transmitting light, reflecting light, and absorbing light. With some goals, pursuit of perfection will lead to simple designs; with others, it will lead to design problems beyond any hope of solving.

Designing the best possible switching component for a computer may prove fairly easy; designing the best possible computer will be vastly more complex. Indeed, what we consider to be "the best possible" will depend on many factors, including the costs of matter, energy and time - and on what we want to compute. In any engineering project, what we call "better" depends on indefinitely many factors, including ill-defined and changing human wants. What is more, even where "better" is well defined, the cost of seeking the final increment of improvement that separates the best from the merely excellent may not be worth paying. We can ignore all such issues of complexity and design cost, though, when considering whether limits actually exist. To define a limit, one must choose a direction, a scale of quality. With that direction defining "better," there will definitely be a best.

The arrangement of atoms determines the properties of hardware, and according to quantum mechanics, the number of possible arrangements is finite - more than just astronomically large, yet not infinite. It follows mathematically that, given a clear goal, some one of these arrangements must be best, or tied for best. As in chess, the limited number of pieces and spaces limits the arrangements and hence the possibilities. In both chess and engineering, though, the variety possible within those limits is inexhaustible. Just knowing the fundamental laws of matter isn't enough to tell us exactly where all the limits lie. We still must face the complexities of design.

Our knowledge of some limits remains loose: "We know only that the limit lies between here (a few paces away) and there (that spot near the horizon)." Assemblers will open the way to the limits, wherever they are, and automated engineering systems will speed progress along the road. The absolute best will often prove elusive, but the runners-up will often be nearly as good. As we approach genuine limits, our abilities will, in ever more areas of technology, cease growing. Advances in these fields will stop not merely for a decade or a century, but permanently. Some may balk at the word "permanently," thinking "No improvements in a thousand years? In a million years? This must be an overstatement." Yet where we reach true physical limits, we will go no further.

The rules of the game are built into the structure of the vacuum, into the structure of the universe. No rearranging of atoms, no clashing of particles, no legislation or chanting or stomping will move natural limits one whit. We may misjudge the limits today, but wherever the real limits lie, there they will remain. This look at natural law shows limits to the quality of things. But we also face limits to quantity, set not only by natural law but by the way that matter and energy are arranged in the universe as we happen to find it. The authors of The Limits to Growth, like so many others, attempted to describe these limits without first examining the limits to technology. This gave misleading results.

Recently, some authors have described the accumulation of waste heat and disorder as ultimate limits to human activity. In The Lean Years - Politics in the Age of Scarcity, Richard Barnet writes: "It is ironical that the rediscovery of limits coincides with two of the most audacious technological feats in human history. One is genetic engineering, the sudden glimpse of a power to shape the very stuff of life.

The other is the colonisation of space. These breakthroughs encourage fantasies of power, but they do not break the ecological straightjacket known as the Second Law of Thermodynamics: Ever greater consumption of energy produces ever greater quantities of heat, which never disappear, but must be counted as a permanent energy cost. Since accumulation of heat can cause ecological catastrophe, these costs limit man's adventure in space as surely as on earth." Entropy is a standard scientific measure of waste heat and disorder. Whenever activities consume useful energy, they produce entropy; the entropy of the world therefore increases steadily and irreversibly. Ultimately, the dissipation of useful energy will destroy the basis of life.

As Rifkin says, this idea may seem too depressing to consider, but he argues that we must face the terrible facts about entropy, humanity, and the Earth. Barnet writes that accumulating heat is a permanent energy cost, limiting human action. Rifkin states that "pollution is the sum total of all of the available energy in the world that has been transformed into unavailable energy." This unavailable energy is chiefly low-temperature waste heat, the sort that makes television sets get warm. But does heat really accumulate, as Barnet fears? If so, then the Earth must be growing steadily hotter, minute by minute and year by year. We should be roasting now, if our ancestors weren't frozen solid.

Somehow, though, continents manage to get cold at night, and colder yet during the winter. During ice ages, the whole Earth cools off. Rifkin takes another tack. He states that "the fixed endowment of terrestrial matter that makes up the earth's crust is constantly dissipating. Mountains are wearing down and topsoil is being blown away with each passing second." By "blown away" Rifkin doesn't mean blown into space or blown out of existence; he just means that the mountain's atoms have become jumbled together with others. Yet this process, he argues, means our doom. The jumbling of atoms makes them "unavailable matter", as a consequence of the "fourth law of thermodynamics," propounded by economist Nicholas Georgescu-Roegen: "In a closed system, the material entropy must ultimately reach a maximum," or (equivalently) that "unavailable matter cannot be recycled," Rifkin declares that the Earth is a closed system, exchanging energy but not matter with its surroundings, and that therefore "here on earth material entropy is continually increasing and must ultimately reach a maximum," making Earth's life falter and die.

A grim situation indeed - the Earth has been degenerating for billions of years. Surely the end must be near! But can this really be true? As life developed, it brought more order to Earth, not less; the formation of ore deposits did the same. The idea that Earth has degenerated seems peculiar at best. Besides, since matter and energy are essentially the same, how can a valid law single out something called "material entropy" in the first place? Rifkin presents perfume spreading from a bottle into the air in a room as an example of "dissipating matter," of material entropy increasing - of matter becoming "unavailable." The spread of salt into water in a bottle will serve equally well.

Consider, then, a test of the "fourth law of thermodynamics" in the Salt-Water Bottle Experiment: Imagine a bottle having a bottom with a partition, dividing it into two basins. In one sits salt, in the other sits water. A cork plugs the bottle's neck: this closes the system and makes the so-called fourth law of thermodynamics apply. The bottle's contents are in an organised state: their material entropy is not at a maximum - yet. Now, pick up the bottle and shake it. Slosh the water into the other basin, swirl it around, dissolve the salt, increase the entropy-go wild! In such a closed system, the "fourth law of thermodynamics" says that this increase in the material entropy should be permanent.

All of Rifkin's alarums about the steady, inevitable increase of Earth's entropy rest on this principle. To see if there is any basis for Rifkin's new worldview, take the bottle and tip it, draining the salty water into one basin. This should make no difference, since the system remains closed. Now set the bottle upright, placing the saltwater side in sunlight and the empty side in shade. Light shines in and heat leaks out, but the system remains as closed as the Earth itself. But watch - the sunlight evaporates water, which condenses in the shade! Fresh water slowly fills the empty basin, leaving the salt behind. Sunlight brings energy from space; heat radiated back into space carries away entropy (of which there is only one kind).

Therefore, entropy can decline in a closed system and flowers can bloom on Earth for age upon age. Rifkin is right in saying that "it's possible to reverse the entropy process in an isolated time and place, but only by using up energy in the process and thus increasing the overall entropy of the environment." But both Rifkin and Barnet make the same mistake: when they write of the environment, they imply the Earth - but the law applies to the environment as a whole, and that whole is the universe. In effect, Rifkin and Barnet ignore both the light of the Sun and the cold black of the night sky. According to Rifkin, his ideas destroy the notion of history as progress, transcending the modern worldview. He calls for sacrifice, stating that "no Third World nation should harbour hopes that it can ever reach the material abundance that has existed in America."

He fears panic and bloodshed. Rifkin finishes by informing us that "the Entropy Law answers the central question that every culture throughout history has grappled with: How should human beings behave in the world?" His answer? "The ultimate moral imperative, then, is to waste as little energy as possible." This would seem to mean that we must save as much energy as possible, seeking to eliminate waste. But what is the greatest nearby energy waster? Why, the Sun, of course - it wastes energy trillions of times faster than we humans do. If taken seriously, it seems that Rifkin's ultimate moral imperative therefore urges: "Put out the Sun!" This silly consequence should have tipped Rifkin off. He and many others hold views that smack of a pre-Copernican arrogance: they presume that the Earth is the whole world and that what people do is automatically of cosmic importance.

There is a genuine entropy law, of course: the second law of thermodynamics. Unlike the bogus "fourth law," it is described in textbooks and used by engineers. It will indeed limit what we do. Human activity will generate heat, and Earth's limited ability to radiate heat will set a firm limit to the amount of Earth-based industrial activity. Likewise, we will need winglike panels to radiate waste heat from our starships.

Finally, the entropy law will - at the far end of an immensity of time - bring the downfall of the universe as we know it, limiting the lifespan of life itself. The entropy threat is an example of blatant nonsense, yet its inventors and promoters aren't laughed off the public stage. Imagine a thousand, a million similar distortions - some subtle, some brazen, but all warping the public's understanding of the world.

Now imagine a group of democratic nations suffering from an infestation of such memes while attempting to cope with an era of accelerating technological revolution. We have a real problem. To make our survival more likely, we will need better ways to weed our memes, to make room for sound understanding to grow.

Solar Energy and Resources

Natural law limits the quality of technology, but within these limits we will use replicating assemblers to produce superior spacecraft. With them, we will open space wide and deep. Today Earth has begun to seem small, arousing concerns that we may deplete its resources. Yet the energy we use totals less than 1/10,000 of the solar energy striking Earth; we worry not about the supply of energy as such, but about the supply of convenient gas and oil.

Our mines barely scratch the surface of the globe; we worry not about the sheer quantity of resources, but about their convenience and cost. When we develop pollution-free nanomachines to gather solar energy and resources, Earth will be able to support a civilisation far larger and wealthier than any yet seen, yet suffer less harm than we inflict today.

The potential of Earth makes the resources we now use seem insignificant by comparison. Yet Earth is but a speck. The asteroidal debris left over from the formation of the planets will provide materials enough to build a thousand times Earth's land area.

The Sun floods the solar system with a billion times the power that reaches Earth. The resources of the solar system are truly vast, makes the resources of Earth seem insignificant by comparison. Yet the solar system is but a speck. The stars that crowd the night sky are suns, and the human eye can see only the closest. Our galaxy holds a hundred billion suns, and many no doubt pour their light on dead planets and asteroids awaiting the touch of life. The resources of our galaxy make even our solar system seem insignificant by comparison. Yet our galaxy is but a speck. Light older than our species shows galaxies beyond ours. The visible universe holds a hundred billion galaxies, each a swarm of billions of suns.

The resources of the visible universe make even our galaxy seem insignificant by comparison. With this we reach the limits of knowledge, if not of resources. The solar system seems answer enough to Earth's limits - and if the rest of the universe remains unclaimed by others, then our prospects for expansion boggle the mind several times over. Does this mean that replicating assemblers and cheap spaceflight will end our resource worries? In a sense, opening space will burst our limits to growth, since we know of no end to the universe. Nevertheless, Malthus was essentially right. In his 1798 Essay on the Principle of Population, Thomas Robert Malthus, an English clergyman, presented the ancestor of all modern limits-to-growth arguments. He noted that freely growing populations tend to double periodically, thus expanding exponentially. This makes sense: since all organisms are descended from successful replicators, they tend to replicate when given a chance.

For the sake of argument, Malthus assumed that resources - the food supply - could increase by a fixed amount per year. Since mathematics shows that any fixed rate of exponential growth will eventually outstrip any fixed rate of linear growth, Malthus argued that population growth, if unchecked, would eventually outrun food production. Malthus failed to predict when limits would pinch us chiefly because he failed to anticipate breakthroughs in farm equipment, crop genetics, and fertilisers.

Some people now note that exponential growth will overrun the fixed stock of Earth's resources, a simpler argument than the one Malthus made. Though space technology will break this limit, it will not break all limits. Even if the universe were infinitely large, we still could not travel infinitely fast. The laws of nature will limit the rate of growth: Earth's life will spread no faster than light. Steady expansion will open new resources at a rate that will increase as the frontier spreads deeper and wider into space. This will result not in linear growth, but in cubic growth. Yet Malthus was essentially right: exponential growth will outrun cubic growth as easily as it would linear. Calculations show that unchecked population growth, with or without long life, would overrun available resources in about one or two thousand years at most. Unlimited exponential growth remains a fantasy, even in space. Do other civilisations already own the resources of the universe? If so, then they would represent a limit to growth. The facts about evolution and technological limits shed useful light on this question.

Since many Sunlike stellar systems are many hundreds of millions of years older than our solar system, some civilisations should be many hundreds of millions of years ahead of ours. We would expect at least some of these civilisations to do what all known life has done: spread as far as it can. Earth is green not just in the oceans where life began, but on shores, hills, and mountains. Green plants have now spread to stations in orbit; if we prosper, Earth's plants will spread to the stars. Organisms spread as far as they can, then a bit farther.

Some fail and die, but the successful survive and spread farther yet. Settlers bound for America sailed and sank, and landed and starved, but some survived to found new nations. Organisms everywhere will feel the pressures that Malthus described, because they will have evolved to survive and spread; genes and memes both push in the same direction. If extraterrestrial civilisations exist, and if even a small fraction were to behave as all life on Earth does, then they should by now have spread across space.

Like us, they would tend to evolve technologies that approach the limits set by natural law. They would learn how to travel near the speed of light, and competition or sheer curiosity would drive some to do so. Indeed, only highly organised, highly stable societies could restrain competitive pressures well enough to avoid exploding outward at near the speed of light. By now, after hundreds of millions of years, even widely scattered civilisations would have spread far enough to meet each other, dividing all of space among them. If these civilisations are indeed everywhere, then they have shown great restraint and hidden themselves well. They would have controlled the resources of whole galaxies for many millions of years, and faced limits to growth on a cosmic scale.

An advanced civilisation pushing its ecological limits would, almost by definition, not waste both matter and energy. Yet we see such waste in all directions, as far as we can see spiral galaxies: their spiral arms hold dust clouds made of wasted matter, backlit by wasted starlight. If such advanced civilisations existed, then our solar system would lie in the realm of one of them. If so, then it would now be their move - we could do nothing to threaten them, and they could study us as they pleased, with or without our cooperation. Sensible people would listen if they firmly stated a demand. But if they exist, they must be hiding themselves - and keeping any local laws secret.

The idea that humanity is alone in the visible universe is consistent with what we see in the sky and with what we know about the origin of life. No bashful aliens are needed to explain the facts. Some say that since there are so many stars, there must surely be other civilisations among them. But there are fewer stars in the visible universe than there are molecules in a glass of water. Just as a glass of water need not contain every possible chemical (even downstream from a chemical plant), so other stars need not harbour civilisations. We know that competing replicators tend to expand toward their ecological limits, and that resources are nonetheless wasted throughout the universe. We have received no envoy from the stars, and we apparently lack even a tolerably humane zookeeper.

There may well be no one there. If they do not exist, then we need not consider them in our plans. If they do exist, then they will overrule our plans according to their own inscrutable wishes, and there seems no way to prepare for the possibility. Thus for now, and perhaps forever, we can make plans for our future without concern for limits imposed by other civilisations.

Economic Impacts

Nanotechnology is the creation and utilisation of materials, devices, and systems through the control of matter on the nanometer-length scale, that is, at the level of atoms, molecules, and supramolecular structures. The essence of nanotechnology is the ability to work at these levels to generate larger structures with fundamentally new molecular organisation. These "nanostructures, " made with building blocks understood from first principles, are the smallest human-made objects, and they exhibit novel physical, chemical, and biological properties and phenomena.

The aim of nanotechnology is to learn to exploit these properties and efficiently manufacture and employ the structures. Control of matter on the nanoscale already plays an important role in scientific disciplines as diverse as physics, chemistry, materials science, biology, medicine, engineering, and computer simulation. Nanotechnology is fundamentally changing the way materials and devices will be produced in the future. The ability to synthesise nanoscale building blocks with precisely controlled size and composition and then to assemble them into larger structures with unique properties and functions will revolutionise segments of the materials manufacturing industry.

Nanostructuring is expected to bring about lighter, stronger, and programmable materials; reductions in life-cycle costs through lower failure rates; innovative devices based on new principles and architectures; and use of molecular/cluster manufacturing. Molecular/cluster manufacturing takes advantage of assembly at the nanoscale level for a given purpose. Structures not previously observed in nature can be developed. Challenges include synthesis of materials by design, development of bio- and bioinspired materials, development of cost-effective and scalable production techniques, and determination of the nanoscale initiators of materials failure. Applications of nanotechnology to materials and manufacturing include the following:

— Forming nanostructured metals and ceramics at exact shapes without machining

— Improved colour printing brought about by nanometer-scale particles that have the best properties of both dyes and pigments

— Nanoscale cemented and plated carbide materials and nanocoatings for cutting tools, and other electronic, chemical, and structural applications

— New standards for measurements at nanoscale
— Nanofabrication on a chip with high levels of complexity and functionality

Nanoelectronics

The Semiconductor Industry Association (SIA) has developed a roadmap for continued enhancements in miniaturisation, speed, and power reduction in information processing devices, e. g. sensors for signal acquisition, processors, memories and displays. The SIA's 1997 edition of The National Technology Roadmap for Semi conductors projects the advances required in all the industries that support semiconductor manufacturing to maintain the historical rate of improvement (Moore's Law) of integrated circuits. The projections extend to the year 2012, at which time the smallest component of a device would have a linear dimension of 50 nm. However, for years beyond 2006 and device features 100 nm or smaller, the roadmap is filled with the notation "No Known Solution. " Indeed, in the Sept. 24 edition of Science, Paul Packan of Intel described the technical difficulties currently experienced in semiconductor manufacturing. He stated that Moore's Law "now seems to be in serious danger" and that maintaining the rate of improvement in the next decade "will be the most difficult challenge the semiconductor industry has ever faced. "

The SIA roadmap ends just short of true nanostructure devices because the principles, fabrication methods, and way of integrating devices into systems are unknown. The roadmap explicitly states that "sustained government support of semiconductor research is mandatory if this industry is to continue to provide for strong economic growth in the U. S. " and recognises that new architectures, materials, and processes will be required to meet the goal of achieving 100 nm feature sizes.

The lead time for science maturing into technology is approximately 10 to 15 years; now is the critical time for Government investment in the science and technology of nanostructures for timely impact in information technology. Further, the investment will have spin-offs that enable the attainment (or acceleration) of other SIA roadmap goals. The area of magnetic information storage is illustrative. Other potential nanoelectronics and computer technology breakthroughs include the following:

— Nanostructured microprocessor devices that continue the trend in declining energy use and cost per gate, thereby potentially improving the efficiency of computers by a factor of millions

— Higher transmission frequencies and more efficient utilisation of the optical spectrum to provide at least ten times more bandwidth, with consequences in business, education, entertainment, and defence

— Small mass storage devices with capacities at multi-terabit levels, a thousand times better than today

— Integrated nanosensor systems capable of collecting, processing, and communicating massive amounts of data with minimal size, weight, and power consumption

Other potential applications of nanotechnology include affordable virtual reality stations to provide individualised teaching aids (and entertainment); computational capability sufficient to enable unmanned combat and civilian vehicles; and communication capability that obviates much commuting and other business travel in an era of increasingly expensive transport fuels.

The stringent fuel constraints for lifting payloads into earth orbit and beyond, and the desire to send spacecraft away from the sun (diminishing solar power) for extended missions, compel continued reduction in size, weight, and power consumption of payloads. Nanostructured materials and devices promise solutions to these challenges. Nanostructuring is also critical to the design and manufacture of lightweight, high strength, thermally stable materials for aircraft, rockets, space stations, and planetary/solar exploratory platforms. Moreover, the low gravity, high vacuum space environment may aid development of nanostructures and nanoscale systems that cannot be created on Earth. Applications of nanotechnology in this area are broad, with potential relevance to other fields as well:

- Low-power, radiation-tolerant, high-performance computers
- Nano instrumentation for micro spacecraft
- Avionics made possible by nanostructured sensors and nanoelectronics
- Thermal barrier and wear-resistant nanostructured coatings

Environment and Energy

Nanotechnology has the potential to significantly impact energy efficiency, storage, and production. It can be used to monitor and remediate environmental problems; curb emissions from a wide range of sources; and develop new, "green" processing technologies that minimise the generation of undesirable by-product effluents.

The impact on industrial control, manufacturing, and processing will be impressive and result in energy savings. Several technologies that utilise the power of nanostructuring, but that were developed without benefit of the new nanoscale analytical capabilities, illustrate this potential:

- A long-term research programme in the chemical industry on the use of crystalline materials as catalyst supports has yielded catalysts with well-defined pore sizes in the range of 1 nm; their use is now the basis of an industry that exceeds $30 billion/year.

— The discovery of the ordered mesoporous material MCM-41 produced by Mobil Oil Co., with pore size in the range 10 to 100 nm, is now widely applied in removal of ultrafine contaminants.

— Several chemical manufacturing companies are developing a nanoparticle-reinforced polymeric material that can replace structural metallic components in the auto industry. Widespread use of those nanocomposites could lead to a reduction of 1. 5 billion liters of gasoline consumption over the life of one year's fleet of vehicles and reduce related carbon dioxide emissions by more than five billion kilograms annually.

— The replacement of carbon black in tires by nanometer-scale particles of inorganic clays and polymers is a new technology that is leading to the production of environmentally friendly, wear-resistant tires.

Potential future breakthroughs also include use of nanostructured materials for environmental and nuclear waste management.

Global Trade

Technology is the major driving factor for growth at every level of the U. S. economy. Nanotechnology is expected to be pervasive in its applications across nearly all technologies. Investment in nanotechnology research and development is necessary to maintain and improve our position in the world marketplace. The proposed nanotechnology initiative will allow the development of critical enabling technologies with broad commercial potential, such as nanoelectronics, nanostructured materials and nanoscale-based manufacturing processes. These are necessary for U. S. industry to take advantage of nanotechnology innovations and improve our capability to compete globally.

Fundamental Discoveries

The potential indicated above for nanotechnology to transform so many aspects of human existence is almost without precedent. In the last few years, applying fundamental discoveries related to nano-technologies has already developed multibillion-dollar businesses. These latter include giant magneto resistance (for hard disks), nanolayers (for data storage and the photographic industry), nanoparticles (for drugs in the pharmaceutical field and colorants in printing), confinement effects (for optoelectronic devices and lasers), nanostructured materials (for nanocomposites and nanophase metals), and chemical and biological detection (for national security and the food industry). Fundamentally novel phenomena and processes have led to new, high-value-added technologies. Investment in enabling basic research and infrastructure for nanoscience and engineering promises extraordinarily high economic and societal returns; it is due primarily to this fact that the need to establish a nationally coordinated nanotechnology initiative is so compelling. However, 1999 IWGN workshop contributors concluded that a much greater investment could be utilised effectively to increase the rate of discovery, and in fact, many opportunities are not being pursued because of lack of resources. Currently, only about one-third of high quality academic research proposals are being funded.

Doubling current Federal expenditures in fiscal year 2001 would ensure that more of the best ideas are funded, increasing the current rate of scientific breakthroughs and drawing more strong researchers to enrich the field. Private industry cannot be expected to fund advancements in basic knowledge on a significant scale. Once nanotechnology has been firmly established, the Government investment will be dwarfed by industry R&D investment, which in the high-technology areas generally is about 10% of sales. Until that time, Government agencies should stimulate and support basic research and infrastructures that will enable subsequent development and commercialisation.

Nanoscience research in the United States has developed in open competition with existing disciplines. This has been healthy for the early stages of development, but it is also the main reason that U. S. nanotechnology research efforts tend to be fragmented and overlap among areas of relevance and sources of funding. A coordinated effort should focus resources on enabling nanoscience and engineering, on developing infrastructure, stimulating cooperation, and avoiding unwanted duplication of efforts. It should take full advantage of the extraordinarily rich research opportunities and potential technological advances promised by early nanoscience work.

A key feature of the IWGN proposal for a national nanotechnology initiative is promotion of synergistic efforts in research, development, and education among Federal agencies. It is the consensus of the IWGN•participants and contributors that the promises of nanotechnology can best be realised through long term and balanced investment in U. S. infrastructure and human resources in five R&D categories in particular:

— *Nanostructure properties:* Investigate biological, chemical, electronic, magnetic, optical, and structural properties in nanostructures.

— *Synthesis and processing:* Enable atomic and molecular control of material building blocks to provide the means to assemble and utilise these tailored building blocks for new processes and devices in a wide variety of applications. Extend the traditional approaches to patterning and microfabrication to include parallel processing with proximal probes, stamping, and embossing. Give particular attention to the interface with bionano structures and bio-inspired structures, to multifunctional and adaptive nanostructures, to scaling approaches, and to affordability at commercial scales.

— *Characterisation and manipulation:* Develop new experimental tools to broaden the capability to measure and control nanostructured matter, including developing new standards of measurement. Pay particular attention to tools capable of measuring and/ or manipulating single macro- and supra-molecules of biological interest.

— *Modelling and simulation:* Accelerate the application of novel concepts and high performance computation to the prediction of nanostructured properties, phenomena, and processes.

— *Device and system concepts:* Stimulate the innovative application of nanostructure properties to new technologies.

The United States does not dominate nanotechnology research. There is strong international interest, with nearly twice as much ongoing research overseas as here. Other regions, particularly Japan and Europe, are supporting work equal to the quality and breadth of the science done in the United States, because there too, scientists and national leaders have determined that nanotechnology has the potential to be a major economic factor during the next several decades. This situation is unlike other post-World War II technological revolutions, where the United States enjoyed earlier leads.

Since it will be impossible to lead in every aspect of this emerging super-field, the United States should look to partner with other countries through mutually beneficial information sharing, cooperative research, and study by young U. S. researchers at foreign centres of excellence.

5

Nanotechnology Applications in the Energy Sector

It is here that nanoscience and nanotechnology must play a pivotal role. At the root of the opportunities provided by nanoscience to impact our energy security is the fact that all the elementary steps of energy conversion take place on the nanoscale. Thus, the development of new nanoscale materials, as well as the methods to characterise, manipulate and assemble them, creates an entirely new paradigm for developing new and revolutionary energy technologies.

Catalysis controls the rates at which chemical bonds are formed and broken, thereby controlling the yields of desired products over undesired ones in chemical reactions. Catalysis is therefore central to energy conversion and environmental protection—as well as a healthy economy. Conversion of petroleum and other fossil fuels and the cleaner emissions of modern cars and fossil-fuel power plants are made possible by catalysts. Today, we face new challenges in creating alternate fuels, cleaning the environment, dealing with the causes of global warming, and keeping us safe from toxic substances and infectious agents.

Catalysis can help us meet these challenges, but the complexity and diversity of catalysts demand a revolution in the way they are designed and used. This revolution can become reality by the application of new materials and methods emerging from nanoscience.

The opportunities to understand and predict how catalysts work at a fundamental level have unfolded dramatically in the past twenty years with the emergence of powerful predictive methods such as density functional theory; incisive instruments for imaging at the nanoscale; and dedicated facilities for examining catalysts as they function, even at the high temperatures and pressures used in industrial processes. New approaches such as rapid-throughput catalyst testing methods and terascale computational methodologies coupled with the emergence of nanoscience and the research facilities to support it all converge on catalysis science to place it on the threshold of new understanding and new technology for energy conversion.

The research challenge in nanoscience for catalysis is learning to tune the energy landscape of the chemical reactants as they interact with nanostructured catalytic materials. Drawing from the lessons of biology, we must design nanostructured materials to match the structural conformation of the reactant and to control the reaction pathway to the desired product. To accomplish this, new and efficient methods of in-situ characterisation and rapid throughput testing of catalytic properties will be required.

The choice of materials, structural parameters and experimental design must be guided by a continually improving fundamental understanding of the structure-function relationships of nanostructured catalysts. In the foreseeable future, the major source of energy for the nation is found in chemical bonds. Accessing and using these energy sources will rely on our development of efficient, selective, and economical catalysts. Nanomaterials offer a new paradigm for the development of catalysts. Materials that in the bulk form have little or no catalytic activity can exhibit exceptional catalytic behaviour in nanoscale form. The best example is the system of gold nanoparticles supported on a titanium-dioxide substrate. Gold in its bulk form is known for its inert properties, making it useful in corrosion resistant coatings and art and dentalfillings. However, Haruta and co-workers discovered that nanoparticles of gold in the 5 nm regime dispersed over a titania substrate are remarkably active for important reactions such as hydrocarbon epoxidation and CO oxidation at low temperatures.

This initial work has spurred further investigations into the importance of the gold nanoparticle size and the role of the substrate support. However, these materials are typically discovered by Edisonian methods and have complex, nonuniform, and poorly understood nanostructures. If scientists and engineers could precisely design the structures of such materials, they could control the relationship between catalyst structure and catalytic chemistry.

The effects on energy production and use in both domestic and industrial applications would be profound. Nanoscience provides the opportunity for this precise design. Nature's catalysts, enzymes, show how amazingly efficient catalysts can be—they make possible essentially all biological reactions—and, by comparison with today's technology, they show how much opportunity there is for improvement.

Control and efficiency of chemical conversion comparable to that demonstrated by nature are in sight for countless known and unknown reactions, as we learn how catalysts work and how to engineer them for specific reactions. We are faced with a grand challenge of designing completely new catalysts and catalytic processes for as yet unimagined purposes and benefits. The emerging tools of nanoscience will be central to meeting this challenge. With these tools, we can envision such technologies as one-step conversion of cellulose or CO_2 into liquid fuels, economical photocatalytic conversion of water into hydrogen (and oxygen) for a hydrogen economy, and efficient fuel cells for the conversion of hydrogen to power automobiles and heat homes.

Most catalysts need to operate under conditions of high temperature and pressure, so nanomaterials that are stable under these conditions are required. The catalysts also must be stably dispersed—i.e., present on solid supports, where they are accessible to reactant molecules. The supports provide high surface area per unit volume and are therefore porous materials, usually at the nanoscale. Catalysts are usually present on the surfaces of these supports or partially embedded in them. Thus, catalysts are typically nanoparticles within nanoporous materials, and the connections between them are interfaces that are referred to as hidden, because they are not susceptible to imaging as is an extended flat surface.

Catalyst performance is determined by the composition, size, morphology, and degree of uniformity of the nanoparticles, the support, and the interface between them. Thus, to understand catalysts and develop a foundation for predicting their performance, we need to understand the science of nanomaterials and nanomaterial interfaces. The science of catalytic nanomaterials brings its own specific challenges beyond these, some of them associated with the reactive nature of these materials and the presence of hidden interfaces.

One of the specific challenges presented by catalytic nanomaterials is to determine their structures in reactive environments. This is essential because the structures and even the compositions change as the environments change, sometimes in subtle ways that are crucial since small changes in structure and composition often affect reactivity profoundly. Furthermore, because subtle changes in nanostructure and composition can have such great consequences in catalyst performance, we must develop rapid, efficient methods for testing the performance of many candidates en route to finding the best catalyst for an application.

New materials for every component of fuel cells are essential to the widespread introduction of this technology. These are sorely needed to considerably improve performance (faster kinetics, less sensitivity to poisons, etc.) and reliability (durability), as well as lower cost. Improved electrocatalysts are vital for anodes and cathodes, and improved ionically conducting membranes in low-temperature fuel cells (below approx 150°C). More specifically, as new more complex catalyst materials are explored, efficient, generally applicable routes to the formation of nanoparticles with diameters from 3 to 5 nm must be found so that the materials can be routinely and rapidly evaluated in realistic fuel-cell configurations and designs.

An example of state-of-the-art synthesis in an energy-related area involves synthesis of a nanocatalyst for hydrogen oxidation in fuel cells. Using techniques from electrochemistry, researchers at Brookhaven National Laboratory in U.S synthesised ruthenium nanoparticles with a sub-monolayer of platinum atoms on the surface for use as catalysts for hydrogen oxidation in fuel cells for the hydrogen economy. This is attractive for catalytic applications because of the potentially low platinum loading of this catalyst since every platinum atom is on the surface. The synthesis techniques allow for careful control of the nanocatalyst and for production of quantities for fuel-cell testing.

A major challenge for nanocatalysts is the development of catalysts with performance characteristics rivaling those of enzymes. These catalysts are responsible for steering almost all of the reactions in living organisms. They are extremely complex in structure and subtle in function.

Working in dilute aqueous solutions in the presence of mixtures of many kinds of reactive molecules at ambient temperatures, they are highly active and 100% selective, and, by biosynthesis, they are replaced as needed in living organisms. Synthetic catalysts with properties approaching those of enzymes would truly revolutionise the technology of energy conversion.

One would envision waste-free or "green chemistry" in the conversion of biomass into fuels. To this end, nature can provide tremendous guidance and insight, and the future of catalysis by nanomaterials is sure to derive countless inspirations from biology. Meeting the ultimate challenge in predicting new catalysts for new conversions will require major development in our science infrastructure.

The development of methods and facilities is required for fundamental understanding of nanomaterials in reactive environments that will elucidate the compositions, structures, and morphologies of nanoparticles, nanoporous materials, and their combinations, including the interfaces. Specifically, it is essential to image these materials at the atomic scale, to measure spectra of various kinds, and to apply computation and theory, all under the same conditions. This major challenge will require new methods for nanomaterials synthesis, new and improved instruments and facilities, and extensive collaborations between experimentalists and theorists. We must develop methods to measure catalyst performance simultaneously with measurements of images and spectra, and this needs to be done with speed and high efficiency.

Advances are needed for imaging three-dimensional materials and interfaces, and spectroscopic methods need to be adapted to nanoscales. Important directions to follow include the effective use of the emerging capabilities for imaging by high-resolution transmission electron microscopy (TEM) and scanning probe microscopy (SPM), spectroscopic characterisation at light sources and neutron sources as well as in dedicated laboratory equipment, and theoretical representation by advanced computation. This will require key improvements in the facilities as well as advancement of sample handling methods and instrumentation to ensure investigation of the same samples under the same conditions by these complementary techniques. Techniques are needed to allow imaging with simultaneous spatial and temporal resolution of changes in catalyst nanostructures as the catalysts operate so as to demonstrate simultaneously how the performance of the catalyst is linked to the images and spectra as they are recorded.

Advances in nanoscience, facilitated by applications of computation and theory in concert with experiment, will facilitate rapid progress in catalysis. The science will benefit greatly from the emerging ability to make uniform catalysts that can be understood fundamentally, without the complications of non-uniform materials that have a range of catalytic properties. The emerging ability to image, manipulate, and spectroscopically interrogate single catalytic nanoparticles will complement the research with uniform nanoparticles.

Investigations of arrays of nanoparticles will facilitate understanding of the effects of nanoarchitecture, including interactions of nanoparticles and transport of reactants and products in catalytic materials. Advances in femto-second science will allow us to time-resolve structural changes on ultrafast timescales, thus adding new dimensions to understanding primary chemical reactions. Theory can now provide reliable calculations of thermodynamics and dynamics for model catalytic reactions. Combinatorial synthesis and rapid screening methodologies are already providing leads to new catalysts at an unprecedented rate, and the integration of these methods with imaging and spectroscopy of functioning catalysts will propel the field forward. New, powerful techniques such as scanning tunnelling microscopy, synchrotron and neutron sources, terascale computational methodologies, and especially the new nanoscience research centres will all converge on catalysis science to place it on the threshold of new understanding and new technology.

Developments in structural biology are already providing insights into the workings of catalysts, leading to the anticipation that bio-inspired catalysts of the future will provide rapid, selective conversions of dilute reactants under mild conditions, even in aqueous environments, as do enzymes. We can envision new catalysts that will allow economical one-step conversion of methane into methanol or other liquid fuels.

Similarly, cellulosic raw materials or CO_2 might be converted into liquid fuels. Improved fuel cell catalysts may lead the way to economical fuel cells for conversion of H_2 or methanol or even gasoline. Improved catalysts may allow small, lightweight devices in automobiles for on-board conversion of liquid fuels into H_2 for fuel cells, eliminating the need for hydrogen storage.

Use of Interfaces

The use of engineered nanostructures at interfaces has demonstrated a compelling potential for improving energy security based on advances in efficient power handling, low-power electronics, energy harvesting, and efficient energy use in lighting. The most significant research challenge impeding these advances is the need to create interfaces that are tailored at the nanoscale to optimise transport of energy in many forms (electrons, phonons, photons, excitons). The issues related to transport of each form of energy are described separately below.

Electron Transport

Electron transport is an old problem of ubiquitous

significance. Both transport of electricity along conducting lines and use of electricity in appliances and information processing is subject to energy loss—the result is energy wasted in creating unwanted heat.

One of the most exciting successes of nanoscience has been the demonstration of loss-less electron conduction, or quantised conduction. This illustration also raises one of the continuing challenges of nanoscience, which is developing controlled-electron transport across nanoscale interfaces—in this case the nanotube-nanotube interface. Transport of electrons across material interfaces is determined by the quantum-mechanically defined electron energies at each side of the interface, and any chemical reactions and energy barriers between the two sides of the interface, as illustrated in the line drawings of Fig. 1. The nature of electronic transport across such barriers is a classical problem in electronics and has been evaluated extensively, generally in the context of flat 2D interfaces.

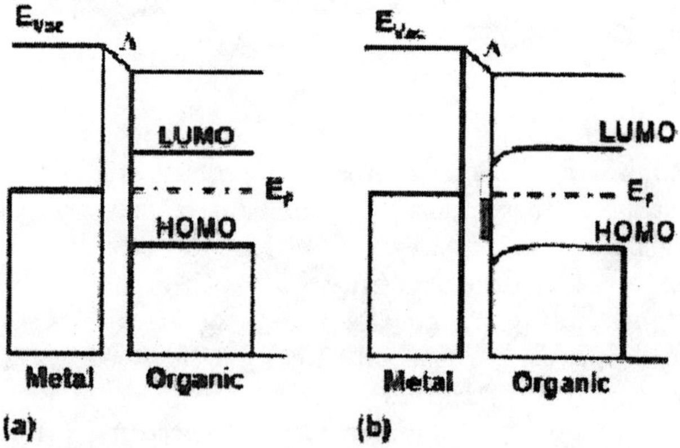

Figure 1. Illustration of some energy issues in electron transport across an interface between a metal and film of organic material.

The new issues that arise in nanoscience are a) the exploration and use of a broad spectrum of non-traditional materials, which introduce a multidimensional parameter space of possible interface chemistry/electronic structure; and b) the complex electric fields that result from the three-dimensional nanoscale structure of interfaces involving nanoparticles and nanodevices.

Exploring the full range of chemistry, atomic structure and three-dimensional configuration that can arise in fabricating nanoscale interfaces introduces a combinatorially large parameter space. The choice of materials and optimising the synthesis/growth/assembly conditions remain a critical and creative parts of the nanoscience endeavor.

Materials may be thin films or membranes and may be patterned via lithographic techniques, imprint techniques, templated self assembly, and/or controlled reactions. Alternatively, materials may be nanoparticles that may be assembled electostatically, via chemical linkages, or via nucleated growth. Interfaces between different combinations of materials may be modulated by control of the structure and chemistry at the interface. The development of nanomaterials systems is coupled with constant evaluation of the desired properties, in this case interfacial transport. Fabricating nanostructures into device configurations is a powerful method for assessing the properties of interfacial transport.

These properties must be understood to enable progress in thermoelectrics, photovoltaics, and lighting, as well as in nanoelectronics. Direct characterisation of the transport properties must be coupled with careful measurements of structure and electromagnetic properties at the interface. The state-of-the-art tools for directly assessing electrical properties at the nanoscale interface are often based on scanned probe techniques or highly focused probe beams (light, electrons). These techniques include the use of magnetic force microscopy and scanning hall probe to measure electron flow around defects, and variants of scanning electric force microscopy and scanning microwave microscopy to measure electric fields and dielectric constants. Ultimately, the use of techniques that rely on spin-resonance detection have the potential for sensing nanoelectronic performance down to the individual spin level.

State-of-the-art experimental research in this area is just beginning to demonstrate direct connections between local atomic structure, local electrical properties, and functional properties of the nanostructure or nanodevice. Theoretical work on the structure and behaviour of materials under the influence of strong fields and the flow of electrical current is also in the exploratory stage. The direct collisions of charge carriers with nanoscale structures and the perturbation of electronic structure via strong local fields both have significant implications for how energy transport at nanoscale interfaces occurs.

Finally, changes in structure under the locally strong fields that are likely to be present in nanodevices has been a well-known physical phenomenon dating from field ion microscopy, and, more recently, in STM induced structure formation. The problems (and potential uses) of these effects in nanomaterials performance must be evaluated using both theoretical and experimental approaches.

Energy Conversion

Thermoelectric materials are ranked by their figure of merit, ZT, which is defined as $ZT = S^2\sigma T/k$, where S is the thermopower or Seebeck coefficient, σ is the electrical conductivity, k is the thermal conductivity, and T is the absolute temperature. Five decades of research have increased room-temperature ZT of bulk semiconductors only marginally, from 0.6 to 1.0. This leads to device performance that is less than 10 percent of the Carnot limit. For the performance to reach 30 percent of the Carnot limit and, thereby, be comparable to their macroscopic gas or vapor-based counterparts, one must develop materials with

$ZT > 3$. The challenge lies in the fact that S, σ, and k are interdependent—changing one alters the others, making optimisation extremely difficult.

The only way to reduce k without affecting S and σ in bulk materials is to use semiconductors of high atomic weight, such as Bi_2Te_3 and its alloys with Sb, Sn and Pb. High atomic weight reduces the speed of sound in the material, and thereby decreases the thermal conductivity. Although in principle there is no limit to increasing ZT, there are no candidate bulk materials in the horizon with $ZT > 3$.

Clearly, this has been one of the grand challenges of materials science, with major implications on energy. Over the last three years, however, there have been reports of dramatic increases in ZT. In all these cases, the material has been found to be nanostructured. This has obviously raised the question of whether quantum effects play any role, since they have been shown to increase ZT. In semiconductors, electrons and holes carry charge, while lattice vibrations or phonons dominate heat transport.

Electrons/holes and phonons have two length scales associated with their transport—wavelength, λ, and mean free path, λ. By nanostructuring semiconductors with sizes comparable to λ sharp edges and peaks in their electronic density of states are produced whose location in energy space depends on size. By matching the peak locations and shape with respect to the Fermi energy, one can tailor the thermopower. Furthermore, such quantum confinement also increases electronic mobility, which could lead to high values of σ. Hence, quantum confinement allows manipulation of $S^2\sigma$ that is difficult to achieve in bulk materials. It is entirely possible, though, that the increase of ZT may be less dependent on quantum confinement of electrons and holes and more on phonon dynamics and transport. For example, if the size of a semiconductor is smaller than the mean free path of phonons and larger than that of electrons/holes, one can reduce thermal conductivity by boundary scattering without affecting electrical transport. While charge transport in thermoelectricity is almost monochromatic (energy levels within a few *kT* around the Fermi energy), heat transport by phonons is broadband over the Brillouin zone. Many bulk thermoelectric materials are alloys because alloy scattering of the short wavelength acoustic phonons suppresses thermal conductivity without significantly altering $S^2\sigma$.

However, the mid-long wavelength phonons remain unaffected and conduct heat in alloys. Can one beat the alloy limit? It is possible that embedded nanostructures in semiconductor alloys could scatter the mid-long wavelength phonons as well due to the similarity in sizes and, thereby, reduce the thermal conductivity below the alloy limit. However, the efficiency of scattering depends on the mismatch in acoustic impedance between the two materials.

Given all these possibilities, it is unclear which effect can be exploited most readily to increase *ZT*. What is clear, though, is that if one has to increase *ZT* beyond 3, it is imperative that we understand how charge and heat transport occur in semiconductor nanostructures, especially across interfaces. These recent discoveries have given scientists hope that it may finally be possible to develop nanostructured thermoelectric materials and devices whose efficiencies are comparable to conventional engines and refrigerators. However, there are many scientific and engineering challenges to be overcome: How do we design interfaces that selectively transmit electrons and holes, while blocking phonons across the whole Brillouin zone?

How can we manufacture bulk quantities of nanostructured thermoelectric materials that have long-term stability at operating temperatures? How can we develop the theoretical and experimental tools required to understand the fundamental science of charge and heat flow at the relevant length and time scales? Answers to these questions could lead to high-performance thermoelectric devices that have the potential to dramatically change the way energy is converted directly between heat and electricity in the future.

Chemistry of Interface

In the long term, solar energy is the only source of renewable energy that has the capacity to fill humanity's technological needs. Major improvements in the conversion efficiency, cost, and stability of solar devices are needed for the increased application of solar energy. As indicated in a recent report, "The grand challenge for solar energy utilisation is to develop conversion systems that are stable and robust for a 20-30 year period and that result in a ten-to-fifty-fold decrease in the cost-to-efficiency ratio for the production of electricity and fuels." While cells based on single-crystal silicon have advanced markedly over the years in terms of increased performance and lower cost, it is doubtful that such cells will be suf.ciently inexpensive to compete with fossil-fuel based electricity and hydrogen.

Nanotechnology has considerable potential to realise cost reduction of photovoltaic cells, but must do so at cell efficiencies greater than 14-15 percent and at costs less than $100/m^2 in order meet a cost goal of $1/W. Nanostructured photovoltaic devices must also be able to endure 15-30 years of outdoor operation with daily cycling to temperatures of 80-100°C at peak insolation. Thin film cells based on nanoparticulate materials can be fabricated and show reasonable efficiencies and stabilities. For example the photovoltaic cell based on dye-sensitised nanoparticles of TiO_2 has been the subject of intense study for several years and can be fabricated in an inexpensive flexible plastic format. The operation of the cell (Fig. 2) is based on photogeneration of electron-hole pairs in the dye, with the electron rapidly moving into the TiO_2 and the hole being filled by a suitable electron donor (e.g., liquid phase electrolyte containing iodide).

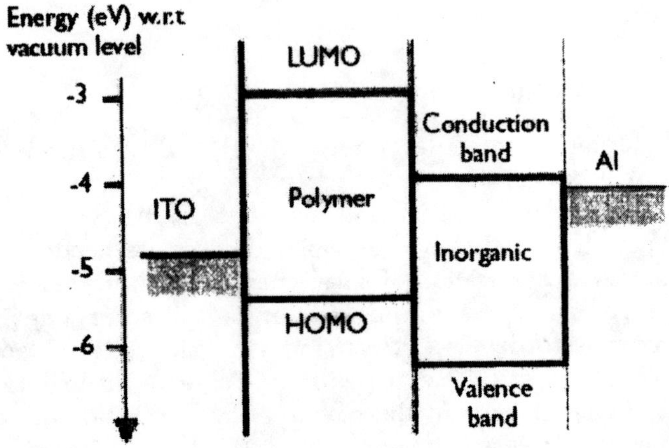

Figure 2. Grätzel cell.

Problems remain in terms of improving the efficiency and stability. The challenge is also to eliminate the liquid phase and to discover alternative electron donors, preferably based on solid-state materials. The liquid phase can be eliminated by making heterojunction polymer devices that contain blends of conjugated polymers (e.g., MEH-PPV) with a nanostructured inorganic material. Photon absorption creates excitons in the polymer, which diffuse about 20 nm before recombining.

Hence, to achieve reasonable efficiencies, the electron-capturing inorganic electrode must be placed within the exciton diffusion lengths. The challenge is to cost-effectively manufacture such composites on a large scale with precision down to 10-20 nm. In addition, hole mobilities in the organic phase must be increased to improve efficiency. This requires fundamental research in charge transport in polymers. Nanostructured materials, such as quantum dots and wells, enable the collection of more than one electron-hole pair per photon.

These nanostructures contain absorber layers for broad spectral absorption collect charge carriers or excitons with enhanced efficiency in low-quality materials. They all have potential for cost and performance breakthroughs in photovoltaics. The challenge again is to cost-effectively manufacture quantum structures on a large scale. Finally, biology has developed effective and durable nano-scale technologies for solar conversion through over a billion years of evolution.

The process of photon-electron energy conversion in molecules is essentially a nanoscale process, which must be integrated with inorganic nanostructured materials for photovoltaics. It makes eminent sense to uncover the basic chemical and physical principles of photosynthesis and apply them to the development of nanoscale technologies for solar conversion.

Light Emitters

The use of semiconductor based light emitters, (composed either of inorganic (LEDs) or organic materials (OLEDs)), for general illumination, is a rapidly developing technology that offers the potential of immense energy savings to the nation and the world within a decade or two. However, a number of science and technology obstacles must be overcome in order for solid-state lighting to reach its potential.

The majority of these obstacles involve understanding electron-hole recombination at nanoscale interfaces. LEDs and OLEDs each have their own potential markets. It is anticipated that OLEDs will serve as large-area, flexible, lightweight sources that will create new lighting opportunities.

The materials employed in OLEDs are thin films of polymers or small molecules that effciently transport injected charge (Fig. 3). Generally, organic light-emitting devices are prepared with discrete layers of different materials for hole and electron transport. In most cases, the emitter material serves a dual purpose as both the emitter and the hole or electron transporter. Further improvement of OLED performance both for display and indoor lighting applications will be advanced through a programme of studying carrier injection, recombination, and light extraction in single crystals.

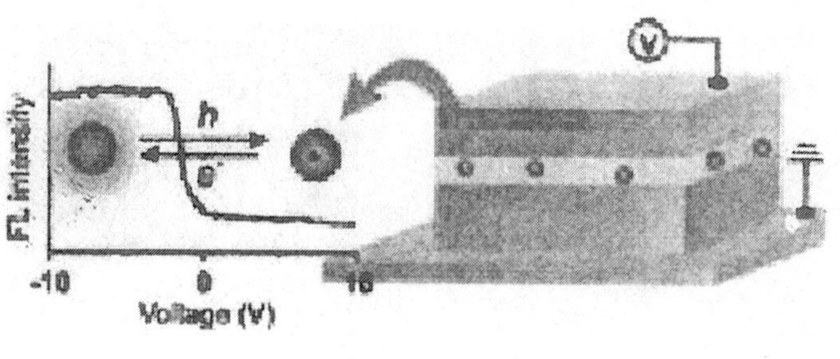

Figure 3. Novel configuration for measuring transport properties relevant to OLEDs.

Such a programme will involve the study of heteroepitaxial interfaces both in the FET configuration as well as in the semiconductor/semiconductor configuration. The microscopic processes governing carrier recombination in single crystal interfaces have yet to be addressed, but clearly represent a fundamental route for OLED development. Moreover, the technology of processing single crystals of the organic material used in OLEDs has recently advanced to permit such a study.

Monochrome LEDs are achieving energy efficiencies as high as 50 percent in the red and on the order of 20-25 percent in the blue LEDs are already 10 times more energy efficient than their filtered incandescent counterparts. Thus they have already replaced over one third of the traffic lights in the country, saving ~\$1000 per intersection per year in electricity.

Achieving acceptable white light for general illumination requires an efficient blue LED, for which the crucial issues are the energy transport at the nanoscale interfaces in phase-separated GaN-based alloys and, specifically, their effects on recombination processes. One of the least understood and technologically most critical issues is a microscopic understanding of the interplay between charge-carrier transport, non-radiative defects, exciton confinement, and oscillator strength across the interfaces in partially phase-separated semiconductor alloys.

Another issue in applications of LEDs is control of lighting colour. Nanocrystals are promising candidates for colour-selectable emitters that combine high (potentially 100%) photoluminescent quantum yields (QY) with chemical flexibility and processibility. Even in the form of a single monolayer they can produce significant power outputs on the order of watts per cm^2 estimated for a nanocrystal packing density of $10^{12}cm^{-2}$. Furthermore, by combining nanocrystals of different sizes, one can easily achieve multicolour operation and white-light emission.

Functions of the Nanoscale

At the foundation of nanoscience are the new phenomena and properties that emerge as materials are constructed at the nanometer scale. These properties originate from the fact that nanometer materials, consisting of only 100-10,000 atoms, have unique structures that cannot be extrapolated from our understanding of traditional bulk materials that we have routinely used in daily life. These properties include, for example, reactivity for more efficient catalytic fuel production (hydrocarbon, H, solar) and for removal of environmental pollutants such as sulphur dioxide, and mechanical strength for ultra-light and strong composite materials.

At the heart of this functionality is the structure of the nanomaterial. Understanding this relationship between structure and function will allow us to develop synthesis techniques that optimise structure and will ultimately lead to the creation of functional nanomaterials by design. A primary difference between nanoscale particles and bulk materials is the fraction of atoms at the surface or interface.

Nearly all the atoms of a nanoparticle are at the surface or interface. This results in atomic composition and structure that are not found in macroscopic materials. Atomic bonds at surfaces and interfaces are distorted as compared to bulk materials. The high fraction of surface atoms leads to segregation phenomena that are highly sensitive to the size of the nanoparticle. This structural uniqueness leads to remarkable physical properties. For example, gold in its bulk form is known to be chemically inert. However, nanoparticles of gold on a titania substrate exhibit remarkable catalytic properties, such as the ability to oxidise the hazardous gas carbon monoxide and decompose sulphur dioxide, a primary source of acid rain and smog. In fact the efficiency of this material is 10 times greater than that of commercial catalysts used in today's automobile catalytic converters for the decomposition of sulphur dioxide.

Similarly, the discovery of nano-structured carbon, such as carbon nanotubes, opens the door to the development of a new class of mechanical and electronic materials. Their structure leads to an unprecedented tensile strength-to-weight ratio and can lead to new composite materials based on these properties.

Furthermore, the detailed carbon structure of the tube (chirality) leads to carbon nanotubes (CNT's) that are conducting or semiconducting. Future applications of these properties include efficient energy transmission and nano-electronics.

These two examples also serve to highlight the range of our understanding of the coupling between structure and function. The detailed atomic and electronic structure of CNTs are known to a large extent. Characterisation of the relationship between the atomic chirality of the tube and its electronic properties is well known and has led to a detailed understanding of the electronic band structure that is derived both by quantum mechanical calculations and experimental measurements.

Present characterisation techniques only allow the measurement of cluster sizes, but not the atomic structure of the cluster or the more important cluster surface. Without this vital information, high-quality, theoretical calculations of chemical reactivity of this system are impossible and advances must rely on intuition or serendipidous discovery. The overarching challenge that we face in designing novel nanomaterials is establishing the physical and chemical principles that relate the changes in molecular structure to the functionality that emerges at nanometer length-scales.

Because we lack these principles, new nanoscale phenomena are being discovered experimentally at an ever-increasing rate that is far outpacing advances in our physical understanding. Effects of quantum confinement, the enhancement of interfacial free energies over bulk free energies, and the importance of surface states have all been recognised since the inception of nanoscience, but a complete theoretical treatment of these factors has yet to emerge. Moreover, many effects of nanoscale structure have only just begun to be explored theoretically, such as fluid flow in nanoporous structures for which the electric double layer is less than the pore size.

One of the discoveries that has emerged from recent research is that the impact of nanoscale structure on function extends beyond individual nanoscale elements to hierarchical architectures. This is because length-dependent factors such as the proximity of interfaces become comparable to the length-scale of electron, exciton, and phonon transport. As a result, dramatic enhancements in properties emerge. Some specific examples include creation of nanocrystal thin-film composites that increase the efficiency of photo-induced energy transfer to as much 100%, nanolaminated solids to depress thermal transport below what was previously called the minimum thermal conductivity, and self-organised arrays of nanoscale structures on alloy films to enhance catalytic reactivity.

At a very fundamental level, structure and function are magnifiers of the evolution in the energetic drivers that occurs as materials dimensions are reduced. This is because the thermodynamic and kinetic factors that determine pathways of materials formation and both the morphology and stability of structures depend exponentially on those drivers. In many respects, it is for this reason that nanoscience is such a rich source of new phenomena. Developing a comprehensive understanding of the physical and chemical principles that lead to these phenomena will remain a major challenge for the foreseeable future.

Theory, modelling, and simulation

Although theory, modelling, and simulation have seen tremendous advances in the last 10 years, nanostructures and their function present challenges that cannot be presently addressed. These fall into two broad categories: (1) high fidelity calculations of atomic and electronic structure and (2) linking this information to predicting functionality. A major contributor to the first challenge is the complexity of nanostructures that derives from their structural non-periodicity and chemical heterogeneity. For the second challenge, the lack of reliable and quantitative measurements makes benchmarking of new techniques and approaches nearly impossible. However, the development of our ability to model and predict the function of nanostructures will be a determining factor in the success of nanotechnology for energy applications.

We are not presently able to calculate accurately the structure of supported nanoclusters on surfaces. The immediate ramification of this is that we do not understand the origin of the catalytic behaviour of the nanoparticle gold/titania system and how it relates to the structure of the particle surface and the interface. Even further from our reach is the ability to calculate the pathways and energy barriers for chemical reactions on these particles. A major goal of our efforts should be to develop theory, modelling, and computational capabilities that will permit the prediction and design of functional nanostructures. This will enable the rational design of new materials for a wide range of important energy applications.

Architectures of Nanostructured Materials

Meeting the needs of energy applications over the next decade will require new classes of nanostructured materials capable of functionalities and efficiencies far beyond what is currently considered state of the art. Research will be needed to provide the linkage between the needed functions and the attainable architectures.

Furthermore, novel strategies for assembly of pre-designed architectures will be required to make possible the fabrication of these architectures, initially in small quantities, but eventually in bulk. In order for nanoscale architecture to provide needed capabilities for energy applications in both the near and long terms, we will need to overcome significant technological barriers. Over the next decade, investment in fundamental studies of the properties of nanostructured materials and the methods for assembly of these materials will fuel development of methods that will allow the following capabilities.

1. Prediction of the nanoscale architectures that will achieve extraordinary and novel material properties for energy applications. These efforts will include design

of materials with novel functionalities that allow us to greatly improve the efficiency of:

— Transport of mass and energy

— Energy storage

— Energy conversion

— Energy production

2. Appropriate control of the architecture of materials at the nanoscale to make possible fabrication of structures predicted to exhibit desirable functions.
3. Development of programmed assembly processes to implement cost-effective fabrication of the structures predicted to exhibit desirable functionalities.
4. Engineering of architectures that integrate nanoscale capabilities into micro and macroscale systems that address energy applications and needs. Full incorporation of nanostructured materials will drive redesign of macro systems to take optimal advantage of these material properties.

Living systems provide a proof of principle that the programmed assembly of nanoscale components can be integrated into complex, functional structures and applied to macroscale systems. Studies of the directed assembly of virus particles have provided some insight into the logic of the processes that result in these viruses.

Although biological systems may have been the origin of the concept of selfassembly, most complex biomolecular systems assemble through processes in which a greater degree of control can be applied. This frequently requires the participation of accessory proteins that guide or catalyse portions of the assembly process. By examining the assembly of viral protein coats and other biological structures, researchers have identified a set of principles that are important to biological systems' ability to programme the assembly of complex architectures. The key principles include:

— Controlled conformational switching (allowing individual components to turn on and off interaction potentials)

— Construction by linear pathways (where the sequence of events is programmed just as we would assemble the axles before adding wheels to a car)

— Assembly from subassemblies (to span multiple length scales)

— Common building blocks to build different structures (contributes to efficiency in synthesis and production)

— The use of templates and jigs for form determination

— Staged assembly (makes sure that subassemblies are added in the appropriate sequence)

Biomolecular systems are capable of exquisite control of assembly proce·ses in which nanoscale particles (e.g., proteins) assemble themselves to form complex, functional, macromolecular machines. Attempts to emulate this process using biomimetic or self-assembling systems have led to the development of numerous functional development including multilayers, two-dimensional arrays of nanoparticles and nanostructured materials. Although many of these have properties important for the application of nanotechnology to energy needs, they fall short of the ultimate goal of complete three-dimensional control of architecture at the molecular level. There are several reasons for this, which can be understood in light of the properties of biological macromolecules.

Complete control of nanoscale architecture starts with the ability to perfectly reproduce a structure—the ultimate goal being massively parallel synthesis using a self-assembly or controlled-assembly process. A necessary condition for construction of identical nanostructures is that the corresponding structural elements making up the structure must be identical to one another.

A structure based on a scaffolding of inorganic nanoparticles suffers from the fact that it is currently impossible to create a population of these particles that are completely identical to one another. Huge numbers of identical biological macromolecules can be readily synthesised, suggesting their potential utility as scaffolds for assembly of a nanoscale structure. A second requirement for control of nanoscale architecture is the ability to design interacting surfaces that bind tightly with high specificity.

Although design of interacting surfaces is beyond the capabilities of molecular biology, it is possible to create combinatorial libraries of molecules from which molecules with the desired binding properties can be selected. This approach works well for the immune system and for a wide variety of biotechnology applications. It is also beginning to work well for selection of biological molecules that bind tightly not only to other biological systems, but to a broad range of inorganic materials. It will soon be possible to design and construct structural elements made up of biological macromolecules that have desired geometry and binding specificity, and assemble them into nanoscale structures that incorporate both organic and inorganic elements.

As an example of this approach, Kischenko et al. used the filamentous bacteriophage M13 as a scaffold for constructing a 30 nm diameter, hollow iron cylinder. Native M13 is a flexible cylinder approximately 6.5 nm in diameter and 1 micrometer in length and is made up largely of 2600 copies of a major coat protein. This protein was chemically modified by covalent attachment of two biotin molecules to each coat protein. Biotin binds non-covalently to a protein, streptavidin, with very high affinity. Consequently, when streptavidin was added to the virus particles, it coated the entire exterior of the particles, creating a particle with a diameter about double that of the original virus particle. Since streptavidin is a tetramer, this construct exhibited many biotin binding sites on its surface, which could be used to add additional structural elements. An iron storage protein, ferritin, was modified by covalent binding of biotin, and the biotinylated ferritin was added to the streptavidin-coated phage particles.

This resulted in a final cylinder about 30 nm in diameter, the outer layer of the cylinder being very rich in iron enclosed in the ferritin molecules. This particular example uses the very high affinity of streptavidin for biotin to achieve linkage of elements that do not normally interact. While these principles provide important clues for programming assembly of multi-functional structures, the principles themselves do not tell us how to design systems that can demonstrate these characteristics.

Currently, much progress is being made in exploring assembly processes in purely synthetic and nano/ bio hybrid systems. Examples include liquid crystals, Languimir-Blodgett and lipid bilayers, quantum dot arrays, lyotropic systems (block co-polymers), and positioning of nanocrystals within protein or DNA frameworks. For example, lyotropic assembly is routinely used to template the structure of mesoporous silica structures that are already used as industrial catalysts.

Assembly in these systems is driven by hydrophilic/ hydrophobic interactions, local mechanical strain gradients, and molecular templating interactions. Even though a theoretical underpinning for these interactions is in many cases well established, our ability to predict the structure and properties that will result from a many-body assembly process remains poorly developed. It is interesting to note that natural systems do not always depend on diffusion to bring the proper nanoscale components into place. In fact, nanoscale machines (motor proteins) are used to transport nanoscale components into their proper place in multiscale architectures. Very recent work is now exploring the use of motor proteins as a means of transporting quantum dots within large-scale architectures.

While this work is in a very early stage and borrows heavily upon the machinery made available in living systems, results are already contributing to our understanding of self-repairing and adaptive-system architectures. Just because nature assembles everything from the nanoscale up does not mean that we should do the same with all materials. Photolithography and other patterning tools can be used to create multi-scale architectures in combination with bottom-up assembly. In fact, the combination of nanoscale assembly within micro or macroscale architectures is likely to dominate the early applications.

The energy efficiency of mechanical machinery can be greatly increased by reducing friction and wear. This problem has dramatic consequences in microscale machinery where interfacial forces dominate mechanical response. Current state of the art for microelectromechanical systems (MEMS) uses an organised molecular layer (about 2 nm thick) to prevent direct adhesion and contact between moving parts. In this case the nano assembly enhances performance of the microscale-device architecture. However, one could image the machine architectures that may result if the multifunctional property of the nano assembly (e.g., modulating surface energy) were used as a design principle.

The ultimate value of multifunctional nanomaterials may be seen in the way that they impact the architecture of the engineering system. Today, a limited number of nanostructured materials exists and only a tiny portion of that multifunctional systems design can be explored. However, nano-inspired system architectures are already developing.

Energy-transfer mechanisms that can be made highly efficient at the nanoscale are leading to new approaches for converting electrical energy into visible lighting. Energy systems inherently demand multiscale architecture. For example, energy conversion often involves charge separation that is best accomplished at the nanoscale. That charge—whether it be for a photovoltaic application, a battery, or a thermal electric—must be collected and distributed at the micro and even macroscale. In the same way, catalysis is inherently nanoscale, involving the detailed position and arrangement of atoms and molecules at the reaction center.

Once again, the demands of moving chemical feedstock and reaction products into and out of the nanoscale region requires organisation on multiple length scales. Nanoscaled materials are proving to exhibit a broad range of properties that are not expected on the basis of material properties in bulk. Exploring the universe of possible nanoscale materials presents a huge challenge because of the vastness of parameter space that could conceivably result in useful new materials. Theoretical approaches to predicting properties of nanoscale materials are needed to predict the architectures that are most likely to provide desired functions. Equally important will be conceptual approaches to shaping the energy landscape involved in the control of the assembly process.

Importance of Infrastructures

Infrastructure to enable the growth of theoretical and experimental research into assembly and architecture and the bio-inorganic interface will be critical to taking advantage of the huge opportunities that exist at this interface. The development of facilities that include resources for both molecular biology and materials research in close proximity will greatly expedite the path forward, both by making possible experiments that require profoundly different sets of tools and by connecting scientists with very different types of expertise.

In addition to the usual array of instrumentation for characterisation of nanostructured materials and particles, it will be important to develop state-of-the-art transmission electron microscopy for exploring the geometry of the nanostructures created. This should be augmented by X-ray and neutron scattering to characterise, as completely as possible, the geometry of the structure created during an assembly process. Computational approaches will require extensive computer power to predict the physical properties of nanostructures and to determine the optimal geometries that should be the goals of the assembly process. Feedback between the experimental determination of physical properties of real structures and the theoretical predictions will help in the optimisation of computer approaches to the prediction of nanostructure properties, aiding in progressive advancement of nanostructure design principles.

Role of TMS

As anticipated in initial planning for the National Nanotechnology Initiative (NNI) in U.S, the combination of theory, modelling, and simulation (TMS) is playing a crucial role throughout nanoscience. As we look to meet nanoscience challenges and enable our future energy security, we have identified the following six TMS needs relevant to the high-impact research directions:

— Provide insight into nanoscale experimental measurements

— Understand function and the impact of dynamics and fluctuations

— Characterise new materials (including catalysts) virtually

— Quantify transport mechanisms (including electron transport) at the nanoscale

— Predict self assembly

— Design and virtually synthesise new materials (including optoelectronic materials)

Today, the TMS capabilities to meet these needs exist in varying degrees of maturity and have been listed approximately in order from most to least mature. Satisfying all of these needs will require substantial investment in fundamental research in theory, simulation methodologies, and algorithms. In particular, among the TMS capabilities for future research investment identified in, four stand out as particularly relevant to the TMS needs for future energy security:

— Methods for bridging temporal and spatial scales

— Many-body electronic structure calculations

— Electron transport including interfaces

— Electronic structure methods for optical properties

Nanotechnology is particularly challenging because to make devices useful at the macroscopic scale, we have to understand how properties at the nanoscale impact behaviour at the macroscale; likewise, we need to know how, during synthesis, manipulation of variables at the macroscale affects nanoscale structure. This interaction between different levels of spatial and temporal detail becomes acute in nanotechnology more routinely than in any other area of computational materials science or computational chemistry. This is especially the case in developing devices (e.g., energy conversion devices) for meeting energy needs, since the scale of the problem suggests large-scale solutions.

Significant challenges exist in incorporating important atomic-scale details into simulations that reach macroscopic length and time scales. Applications that could benefit from such capabilities include chemical reactor design for processing and conversion of fuel-related molecules and for synthesising solid-state lighting, predicting and modelling self-assembly, linking atomic-scale structure of catalysts with reactivity and selectivity, modelling separations, and synthesising materials to realise separations.

Great strides have been made in bridging length and time scales, through both serial coupling of scales, and parallel coupling. In dynamic, solid-state simulations involving rare-event dynamics, advances in capabilities for rapidly detecting transition states and performing on-the-fly kinetic Monte Carlo simulations are promising for achieving long-time simulations in areas such as catalysis, transport in porous materials, and thin-film epitaxy. However, these simulations do not yet reach far beyond the nanometer spatial scale, and their cost can become prohibitive if ab initio potentials are involved.

On the other hand, in simulations of rapid crack propagation in solids, large length scales have been included by coupling simultaneous simulations at the quantum, atomic, and continuum scales. However, these simulations do not extend far beyond the picosecond time scale. To predict and characterise new materials, we must develop the capability to seamlessly bridge both spatial and temporal scales. A second challenge in multiscale modelling for energy applications lies in predicting self-assembly of nanoscale building blocks.

Challenges in this area lie in accurately resolving, through first-principles calculations, the relative contributions of various interactions, including van der Waals dispersion forces, hydrogen bonding, and dynamic structure and chemistry at nanoparticle surfaces. Understanding how these forces affect solvation, depletion, and phase behaviour in seamless, multiscale models is also a challenge. Efficient simulation techniques, such as multi-canonical simulations, need to be developed to reliably predict the phase behaviour of assemblies.

Electronic Structure Calculations

Such calculations are ubiquitous in energy applications, which inevitably involve conversions from one form of matter or energy to another (e.g., catalysis). To realise the full potential of computational modelling and simulation in understanding and controlling chemical change, we must overcome significant theoretical, mathematical, and computational obstacles.

Our simulations must reliably meet specific criteria for both the precision of predictions and their speed of execution. For instance, a factor of 2 to 4 improvement in catalyst efficiency may determine whether a chemical process is economically feasible or not, which at room temperature (25°C) corresponds to 1 kcal/mol accuracy. Only manybody ab initio molecular electronic structure calculations that solve the Schrödinger equation for the electronic wave function can provide this precision.

The TMS challenge is to meet these stringent requirements of precision for realistic models of the chemical systems and their environments, while also allowing for the very large number of calculations actually required to explore structure, dynamics, and statistics. This is not yet possible, and we must extend the progress that has already has been made in reducing the cost of molecular simulations. In addition, current approaches to forming reduced models of large systems (e.g., that embed small quantum regions in larger classical models) are at best heuristic; we must develop reliable techniques for bridging between electronic structure calculations and higher-level methods.

The application of nanoscience to meeting future energy needs will inevitably result in the development of nanoscale devices that transport electrons, particularly across interfaces that may be of quite dissimilar character (e.g., electron transport from a biological molecule to a metal contact across a nano-interface).

At the nanoscale, new electron transport phenomena come into play and classical descriptions become inadequate in the face of quantum phenomena such as tunnelling, fluctuations, confined dimensionality, and the discreteness of the electric charge.

Additionally, nanoscale devices will typically be dominated entirely by surfaces. This presents further complications for electronic materials, which often undergo strong and complex reconstructions. At the theory and simulation level, there does not exist today a consensus on the correct methodology for calculating electron transport in the presence of large electric fields, and the difficulty of the corresponding experiments means that validation of methods is difficult.

An additional problem in electron transport across interfaces is that often the interfaces are not characterised at all, hence the need to resort to computational "construction" of interfaces, i.e., using simulation methods to perform computational nanoscale "experiments" to produce interface structures that should be expected in the experimental systems. Having produced a putative interface, the electron transport is then calculated, recognising that this, in fact, may result in further structural changes to the interface. Hence, the process of developing a model for the interface and performing electron-transport calculations is iterative in nature.

Electronic-structure Methods

Exploiting the optical properties of nanostructures can have significant impact on future energy needs -for example, reducing energy consumption for lighting by replacing current technologies with solid-state lighting or organic light-emitting diodes. At nanoscale dimensions, optical properties of matter are often dramatically altered. For example, while bulk silicon has poor optical properties for optoelectronic devices such as solar cells or lasers, at small dimensions the band gap in silicon can be blue-shifted from the infrared to the optical region; as a result, porous silicon exhibits remarkable room-temperature luminescence. To capitalise properly on such phenomena, a deeper quantitative understanding of the optical properties of matter will be required.

One of the theoretical challenges to understanding optical properties and excitations is that most commonly used methods for structural energies, such as density functional theory, are not well suited for excited-state properties. The problem is exacerbated for nanoscale systems, where the many-body effects are enhanced by the physical confinement of the excitation. New theoretical techniques and computational methodologies focused on optical properties, as well as complementary experimental validation, are required to develop the design tools needed to predict optical properties of nanostructures on a routine basis.

Computational Infrastructure

The diversity of methods used in computational nanoscience, combined with the need for highperformance computing infrastructure to solve realistic problems in nanoscience, pose a considerable challenge. Consequently, many codes have to be optimised on new computer architectures and used in a tightly coupled manner. Interdisciplinary teams need to be formed among scientists that have not typically collaborated and shared codes previously. In view of the many person-years that have to be invested to develop a specific method and computer code, it is quite clear that in order to make computational nanoscience a tool for discovery, algorithm and software developments need to be explicitly coordinated to allow for seamless integration and re-use of methods and the corresponding computer codes. A concerted effort is therefore needed to create community-based software repositories and computational endstations, which are analogous to endstations in large experimental facilities (e.g., a powder-diffraction endstation attached to a large neutron-scattering facility) for nanoscience that will allow scientists to create new simulation methods and respond quickly to the emerging needs of nanoscience.

In such a repository one would, for example, reduce different electronic structure methods to kernels that use a common interface and input/output system. Users as well as other codes that implement a particular aspect of nanoscience, such as electron transport, optical properties, etc., would access these kernels though the common interface. Methods and the corresponding kernels that are best suited for a specific computer architecture can then be selected, optimised, and maintained in a computational endstation. This would allow optimal use of high-performance computing infrastructure while keeping the entire software repository flexible enough to satisfy the changing requirements of theory, modelling, and simulation as the field of nanoscience develops.

Scalable Synthesis

Nanomaterials are important because they can add new functionalities that are simply not present in bulk materials. These may result from (1) high surface area (large surface-to-volume ratio), (2) new intrinsic properties due to small size (quantum size effects) or shape (high aspect ratio and reduced dimensionality), and (3) intimate nanoscale connectivity conferring directionality and minimising losses in energy transport. These new "nano" functionalities can provide electrical and thermal conductivity, field emission, and catalyst support or gas adsorption properties unrivalled by bulk materials. Consequently, the synthesis of functional nanomaterials offers a unique opportunity to make the breakthrough discoveries and truly revolutionary developments that are needed to meet the energy challenge.

The synthesis of functional nanomaterials, that is, materials systems able to deliver greatly enhanced properties at the macro scale where they are needed, entails several interrelated challenges in the quality, quantity, variety, and integrated design and assembly of nanomaterials.

— The quality challenge requires synthesising defect-free nanomaterials with suf.ciently high purity that their ultimate properties can be reliably measured and accurately known. From them, baseline data will be obtained for subsequently controlling synthesis and designing materials systems to deliver their enhanced properties.

— The quantity challenge requires developing, understanding, and optimising readily scalable methods of synthesis for economical, large-scale production of high-quality nanomaterials.

— The variety challenge requires research and development of methods to effciently explore and expand the "library" of enhanced nanomaterials properties that are potentially available. The current library is small and the broad range of potential nanomaterials mostly unexplored.

— The integrated design and assembly (or "nano-to-bulk") challenge requires developing means of synthesis compatible with subsequent (or, better yet, concurrent) assembly of nanomaterials into systems or architectures that exhibit the greatly enhanced nanoscale properties at the macroscale. With these goals, basic research is clearly the central infrastructure need for the synthesis of practical nanomaterials leading to fabrication/processing and system-design techniques that preserve unique nanoscale properties while coupling them to the macroscale. So where do we stand, relative to where we need to be?

Compared with the needs described, the present state of nanomaterials synthesis can be summarised as follows:

— rewarding insights were obtained from limited use of in situ optical and spectroscopic diagnostics during growth processes,

— there is relatively slow feedback from characterisation (properties measurements) and the experimental search for new materials, and

— there is slower feedback from theory and modelling to understand and improve nanomaterials synthesis.

Three key conclusions can be made:

— In situ time-resolved diagnostic methods to understand the synthesis and mechanisms of growth of nanomaterials have proven tremendously valuable. Increased investment specifically for in situ monitoring techniques could have a disproportionately large payoff in understanding and controlling synthesis.

— There is little capability for designing optimised nanomaterials. This deficit is pointed out because fabricating practical materials with enhanced nanoscale properties could soon be assisted by theory, modelling, and simulation, by taking advantage of ongoing advances in computational nanoscience on leadership computers. The parallel development of modelling and simulation (multiscale modelling and virtual synthesis, leading to materials by design), and of time-resolved in situ diagnostics of fabrication/processing techniques, is essential in understanding and controlling the synthesis of complex, natural, engineered nanomaterials systems.

— There is particular need to explore more rapidly the vast space of nanomaterials that are potentially available with new and novel properties. This exploration could be accomplished much more efficiently by development of:

(1) characterisation methods that combine imaging with in situ properties measurements and sample

manipulation, so that the results of synthesis could be evaluated quickly; and

(2) computational methods to guide the search for novel materials, thereby limiting the space that must be searched experimentally.

These recognitions lead to a vision of synthesis as the key, yet probably the limiting, step in what we would like to be a tightly connected loop of synthesis, characterisation, theory/modelling/simulation, and design. This loop can be traversed repeatedly until the objectives in synthesis, quality, quantity, variety, and integrated design have been met. There are two important underlying themes in the energy impact of nanomaterials:

1. It is implicit in many of the BESAC report's research recommendations that degradation of materials in extreme environments (particularly high temperature) is a formidable barrier to developing new energy resources with moderate-to-low initial cost and a long operating lifetime. Consequently, processes must be discovered and systems developed that can operate at low temperatures.
2. There is an opportunity for revolutionary advances through establishing new methods to fabricate materials, including nanoscale self-assembly.

Synthesis of Nanocrystal

Advances in synthetic chemistry have enabled us to make a variety of narrowly dispersed nanocrystals, each small enough to exhibit quantum confinement effects. By tuning the size of the nanocrystal, important physical properties can be tuned, such as optical properties. A recent important advance in synthesis was the demonstration of nanocrystal superlattices. Nanocrystals with specific properties that are brought together will self-assemble to form superlattices, i.e., an ordered array of nanocrystals that spontaneously form under thermodynamically favourable conditions. Using this technique of modular assembly, one can imagine a toolkit from which a variety of nanocrystals with tunable properties can be selected and combined.

One example is that of combining semiconducting nanocrystals (quantum dots) and magnetic nanocrystals, thereby introducing complementary magnetic and optical properties into one material. The magnetic iron oxide nanocrystals are invisible in this region of the spectrum. The superlattice mimics the AB_{13} structure, a known close-packed structure for certain metallic alloys. Scientists at IBM, Columbia University, and the University of New Orleans created this so-called metamaterial.

Using biological approaches enables interesting new directions in synthesising materials relevant to energy issues. One example of this challenge and opportunity is in creating a new approach for artificial photochemical energy conversion based on functionally integrated hybrid (nanoparticle-biomolecule) semiconductor conjugates. Biomolecule-semiconductor conjugates are exceptional in the way that they carry out energy capture and subsequent chemical/electrical conversion. Rajh et.al created a new family of semiconductor-based artificial photosynthetic systems that duplicate the primary light-induced charge separation mechanism in natural photosynthesis. This new class of functional hybrid materials is composed of specifically designed nanoparticles that combine the robustness and photoresponsiveness of inorganic materials with the recognition properties of biological molecules.

Their synthesis exploits the discovery that the surface of metal-oxide nanocrystalline particles differs from that of the respective bulk materials, and is characterised by a highly reactive undercoordinated surface. Using this approach, hybrid conjugates of TiO_2 nanoparticles and DNA oligonucleotides were synthesised, thereby changing the intrinsic properties of the semiconductor and generating new hybrid molecular orbitals, so making the optical properties of metal-oxide nanoparticles tunable throughout the entire visible region.

Additionally, as in natural photosynthesis, these hybrid nanocrystallites display weak electronic coupling of fully separated charges, thus channeling efficient extended charge separation. Insight into charge separation was obtained through measuring the efficiency of photocatalytic silver reduction. The amount of photocatalytic deposition of metallic silver decreases with increasing separation between the TiO_2 surface and secondary GG hole acceptor. This dependence follows an exponential decay with a low attenuating exponent coefficient of 0.045. Consequently, photo-generated holes are transported to very large separation distance of ~80 Å.

This enhanced charge separation is attributed to the fact that upon excitation of TiO_2/dopamine system, charge pairs are instantaneously separated by distances of ~20 Å. This large initial separation diminishes the probability of their recombination, so that holes may be trapped at secondary hole acceptors at very large distances apart.

More Complex Nanomaterials

An area of great opportunity for high impact on the nation's energy security involves expanding the library of functional nanomaterials. For example, the perovskite oxide family of materials exhibits a wide variety of interesting properties in the bulk phase, including high-temperature superconductivity, colossal magnetoresistance, ferroelectricity, and piezoelectric behaviour. Recently, several groups started examining strategies for synthesising this important class of nanomaterials. One example of the state of the art in this field is the recent progress in the nanomaterial synthesis of the ternary perovskite oxides.

Utilising a solid-state synthesis approach, well-dispersed nanocubes of the high dielectric constant oxide

$SrTiO_3$ and nanowires of the ferroelectric $BaTiO_3$ were made. The barium titanate synthesis (left image) generates nanorods of diameter ~50 nm and micrometers long, while the strontium titanate synthesis (right image) results in nanocubes with an edge length of ~80 nm. These syntheses yielded perovskite nanomaterials of very high purity and crystallinity. In addition, this solid-state synthesis approach can be scaled up to produce gram quantities of nanomaterials It also offers promise for a variety of nanomaterials in the perovskite-oxide family.

Highly Parallel Sorting

A major synthesis challenge is to rapidly and efficiently search through materials with new compositions or structures to reveal the relatively few phases with greatly enhanced properties or especially useful new combinations of them. For nanomaterials, the follow-on challenges are to optimise conditions for synthesis and to develop scaled-up methods to produce the quantities needed for energy applications, while maintaining quality and key properties. An example of a synthesis process that can permit rapid screening is continuous compositional spread methods of synthesising films using pulsed laser deposition (PLD-CCS) for complex oxides. By placing an aperture in front of the ablation target, and; combined with precisely controlled synchronisation between the firing of the pulsed laser, exchange of the ablation target, and translation/rotation of the substrate, both the film's thickness and deposition energetics constant across the entire compositional range was maintained.

Moreover, the resulting compositional spread covered a sufficiently large area that the continuously varying film properties could be measured by conventional methods such as SQUID magnetometry, electrical resistivity, and ellipsometry. This PLD-CCS technique was demonstrated by growing a $Sr_xCa_{1-x}RuO_3$ binary alloy with "x" varying from 0 to 1, and with properties ranging from a metallic ferromagnet (x=1) toward antiferromagnetism (x=0).

The versatility of the PLD-CCS method, and its potential for the discovery and optimisation of nanoscale catalysts, recently was demonstrated by applying it to discover the optimum Fe-Mo catalyst composition for growing long dense forests of elongated multi-wall carbon nanotubes. A 2D compositional spread of Fe-Mo catalyst, consisting of "wedges" of Fe and Mo with their thickness increasing orthogonally, was deposited on an Al-coated Si wafer, followed by CVD growth of multi-wall carbon nanotubes. Rapid optical profiling determined the nanotubes' height as a function of position, and SEM verified their continuity.

A clear dependence was found on the catalyst's composition, not simply a threshold amount of one constituent, and nanotubes were grown an order of magnitude longer than previous state-of-the art experiments. These examples demonstrate that non-equilibrium combinatorial search-and-discovery methods such as PLD-CCS reveal broad trends from a single-film growth run, whereas conventional methods require many samples, much more time, and lose precision because of run-to-run irreproducibility. Further development is needed of such economical and high throughput combinatorial and sorting/optimisation methods, both to effciently search for nanomaterials and to scale up their production to the quantities needed for energy applications.

Nanomaterials synthesis techniques offer a path toward new materials that could exhibit a favourable ability for storing hydrogen. Discovered in 1998, single-walled carbon nanohorns (SWNHs) (Fig. 4) are short (~50 nm long), irregular (up to 3 nm in diameter) tubules of carbon, closed at their ends by conical sections. They are easily opened by gentle oxidation and other methods to provide extremely high surface area (1,000 m^2/gram, similar to aerogels) and a rich variety of possible adsorption sites for hydrogen storage.

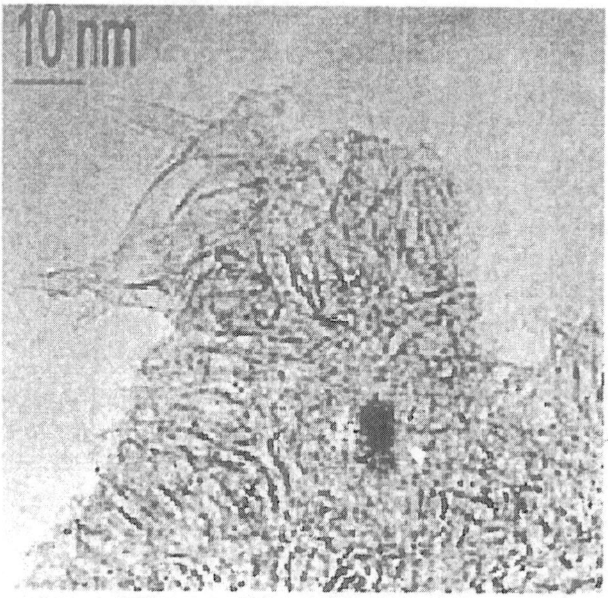

Figure 4. Carbon nanohorn aggregate formed by high-power laser vaporization.

Recent research in Japan demonstrated their success in preventing the aggregation of very small (~2 nm) platinum catalyst nanoparticles, where other forms of nanocarbon fail. SWNHs can be produced either with or without metal catalysts, offering potentially great versatility in processing them for applications. Basic research is needed to explore and optimise their use, both with and without additional catalysts, for hydrogen storage. Such research also will further their potential as catalytic supports for methanol-based fuel cells.

Organic Materials

Electronics made from organic materials offer new routes to

energy-conserving indoor lighting, as well as adaptive sensors for threat reduction. Organic materials derive functionality from their nanometerscale molecular structure. Such molecular texturing imbues these materials with unique attributes, such as mechanical flexibility, molecular diversity, and high electron-density of states.

Organic devices are readily synthesised in polycrystalline form. Their performance is limited, however, by low electron mobility. For instance, mobility is one of the factors restricting luminous efficiency in organic light emitting diodes (OLEDs) made from either polymeric or polycrystalline films. Rapid advances in processing single crystals, on the other hand, promise new routes towards understanding the role of impurities and morphological defects on mobility.

Functional Bulk Materials

Importantly, nanomaterials also play an important role in bulk materials. Nanoscience and nanoscale phases will hugely impact the development of better and entirely new functional bulk materials. Functional bulk materials are essential components in both energy-generating and energy-using systems. Applications ranging from structural components to magnets, filters, and wires provide numerous examples of the benefits of nanoscale precipitates. But the opportunities now open may go well beyond these, to the development of bulk materials that retain a nanograin size yet display at the macroscale level the associated enhanced or new properties.

Even in composites, there is a huge potential for using nanoparticles and nanotubes to produce materials with improved properties and function. The potential impact is wide: materials with longer life in aggressive environments (e.g., high temperatures or radiation), improved energy storage (batteries and fuel cells), functional materials for magnets (since the use of bulk magnets is pervasive in electric motors and generators), and a host of new materials challenges in renewable energy.

Greater energy efficiency will result from introducing materials that can withstand higher temperatures and materials that weigh less. Many technological barriers for nuclear fusion and the next generation of safe fission reactors require developing better materials. Nanoscience could well provide the answers to these difficult technological challenges. Scientifically, there also are many challenges to our understanding. It is accepted that extremely smallgrained materials have much higher strength than anticipated based on grain size alone.

Other unusual properties can likewise be anticipated. However, to realise these properties for actual use within energy systems, it is necessary to preserve the structure and unique properties of nanocrystals and nanoprecipitates, which are often only metastable throughout their fabrication and use. An example of the application of nanoscience to bulk materials is the recent development of a ferritic oxide-dispersive-strengthened (ODS) Fe-12Cr-3W-0.4Ti+0.25Y_2O_3 (wt %) alloy, which contains a secondary, stable, nanophase (less than 4 nm diameter) that dramatically improves its mechanical properties. Atom-probe analyses indicate that these nanoclusters are enriched with Ti, O, and Y compared to the composition of the matrix. Furthermore, these clusters very effectively strengthen the alloy both at room temperature and higher. Long-term tests demonstrate that the nanoclusters reduce the creep rate at 650-900°C by six orders of magnitude compared to conventional tempered martensitic steels. This superior creep resistance represents a major breakthrough in designing ferritic alloys for high-temperature structural applications.

These nanoscale self-assembling clusters are stable and do not coarsen or show significant chemical changes even after 14,000 h at 800°C, or after extended thermal treatments at 1,300°C (0.87 Tm). The alloys used in these initial studies were fabricated by mechanical alloying. To develop controlled synthesis capabilities in these materials, important scientific questions must be addressed: Why are these nanoparticles stable? How do they form? Can the results be duplicated using better controlled fabrication methods? What is the physics that enables these phenomena? Can similar results be obtained for other alloy systems?

The answers to these questions could provide a high technological payoff, including materials for high-radiation environments in fusion and nuclear-energy applications. To enable nanoscale materials to have an impact on the nation's energy security, we must realise an interconnected rapid-feedback loop between synthesis, characterisation, theory, and design. This requires us to provide support to synthesis, materials characterisation, theory, and modelling, and to tie them together. There should be an initiative that will support the science of synthesis in the United States, involving a network of centres of excellence and capability at national laboratories working with distributed synthesis laboratories in universities. This network would quickly make available the basic characterisation tools for materials analysis to synthesis researchers. After promising materials are identified through basic characterisation tools from the synthesis network, there is a need for more advanced investigations to explore the true promise of novel nanomaterials.

This will include advanced neutron-, photon-, and electron-based tools that can probe nanomaterial properties of individual nanomaterials and nanomaterials at work (including, for example, nanocatalysts under high pressures and individual nanotube switching electrons). Finally, we need to undertake the development and validation of predictive computational design capabilities for nanostructured materials.

Quantum simulations can establish microscopic models from crucial first principles to describe and predict the properties of nanomaterials with controllable, quantitative accuracy. The coming-of-age of ab initio theories of interacting many-particles systems, together with the development of related mathematical and computer science tools, will allow simulations of a wide variety of nanostructural properties, thereby opening the possibility of designing optimised materials from first principles and of computing specific properties that can be experimentally validated.

Infrastructure Needs

In order to achieve the level of controlled, scalable synthesis that is required for designed nanomaterials, infrastructural investments are required to support their characterisation and large-scale production, as well as the investigation of "nanomaterials at work."

Using a new generation of neutron-scattering instruments, nanomaterials soon will be characterised in controlled environments that are identical to those used for processing or in energy applications. This will allow important direct investigations of the time and temperature stability of nanocrystalline phases. Insights into the structural evolution of nanophase materials under stress may allow us to understand the origin of the often-observed increased strength with decreasing grain size.

Neutron-scattering studies of deuterated metal samples will yield insights into embrittlement phenomena in nanoscale materials. Realtime studies of the nucleation and growth of nanocrystalline materials from non-crystalline precursors also will be possible, as will nondestructive studies of the precipitation of nanocrystalline phases within bulk materials at high temperatures.

In parallel with these basic studies, a host of technologically important studies also becomes possible.

Among them are quantitative examination of hydrogen and/or hydrocarbon storage in nanocrystalline materials, using neutron diffraction and deuterated samples, and studies of catalysts in real environments to understand both their activity and the technologically important relationships between nanoscale structure and catalyst performance.

6

Nanoelectronics: Opportunities and Challenges

Recent advances in molecular electronics, combined with increased challenges in semiconductor manufacturing, create a new opportunity for çomputer architects - the opportunity to recreate computer architecture from the ground up. New device characteristics require us to rethink the basic abstraction of the transistor. Computer architecture builds computing systems as hierarchies of abstractions. Molecular computing may be a case where a re-examination of the layers of abstraction is required. An alternative computer system architecture, based on dramatically different abstractions:

— Transistor → new molecular devices
— Custom hardware → generic reconfigurable hardware
— Yield control → defect tolerance through reconfiguration
— Synchronous circuits → asynchronous computation
— Microprocessors→hybrid CPU + compiler-synthesised application-specific hardware

Successfully harnessing the power of molecular computing requires us to rethink several key abstractions: the transistor, the circuit, and the ISA. The abstraction of the transistor should be broken down into more basic components, such as controlled signal switching, signal restoration and input/output isolation. The abstraction that a circuit is created at manufacturing time needs to be replaced by the ability to configure circuits at run-time. Finally, the abstraction of an ISA needs to be replaced with a more flexible hardware/software interface: instead of an ISA which hides the details of a processor implementation from the user, the new abstraction will expose more of the underlying computational structure to the compiler, allowing better use of the plentiful resources that become available through molecular computing.

Significant progress has been made in developing molecular scale devices. Molecular scale FETs, negative differential resistors, diodes, and non-volatile switches are among the many devices that have been demonstrated. Advances have also been made in assembling devices and wires into larger circuits. Estimated densities are between 10x and 1000x greater than those predicted by the ITRS 2001 road-map for silicon-based CMOS circuits. In addition to the increases in density, molecular electronics also promises to introduce devices with characteristics not found in silicon-based systems.

One example is the non-volatile programmable switch, which holds its own state without using a memory cell, and can be configured using the signal wires; such a switch has the same area as a wire-crossing. A combination of the restrictions of self-assembly and the small feature size of the devices reduces the likelihood that near-ideal transistors will be available to circuit designers-in fact, three-terminal devices may be completely impractical. Thus, circuit designers will have to use devices that do not each by themselves provide all the features of an ideal transistor. Transistors simultaneously provide the capability of a switch, isolate inputs from outputs and restore (amplify) logical signals. New devices may only provide some of these features. Thus, the propose the *SirM* device model, which explicitly separates the components performing swithing, isolation, restoration, and state storage (memory).

The advantage of this model is that it will make explicit the fact that there are different kinds of devices which serve different roles in a circuit. It will be possible to design a switching network in which signals will degrade and possible interfere with each other. Then between switching networks the designer can insert the proper kinds of devices for signal restoration and input/output isolation. Perhaps the biggest difference between molecular electronics and VLSI is in fabrication technology. Large-scale molecular electronics requires some form of self-assembly.

The use of self-assembly requires not only a different manufacturing process, but requires rethinking every aspect of system design, including, but not limited to: circuit design, architecture, compilation, and testing. When using self-assembly, individual devices and wires are first manufactured, and only later assembled into a circuit. While self-assembly is a very economical process (compared with the cost of today's silicon fabrication equipment), it cannot build patterns of the practically arbitrary complexity that optical lithography creates.

Only simple, crystal-like structures, can be created using self-assembly. Furthermore, defect densities of self-assembled circuits are orders of magnitude higher then in silicon-based devices. Thus, self-assembled circuits and architectures will have to be designed for defect tolerance. In order to implement useful reliable functionality on top of crystal-like structures, post-fabrication customisation is required; this customisation will be used for two purposes (1) to implement the desired functionality and (2) to eliminate the deleterious effects of the defects. Thus, the expect that electronic nanotechnology architectures will be highly reconfigurable.

A reconfigurable architecture allows one to configure a hardware system to implement a particular circuit. Reconfiguration can avoid using the defective components, thus creating a reliable system on an unreliable substrate. One such reconfigurable architecture is the nanoFabric, which is designed to overcome the limitations of self-assembly of molecular scale components.

The basic unit of the nanoFabric is the programmable cross-point. These cross-points are organised into 2-D meshes which themselves are organised into clusters. The fabric is a fine-grained reconfigurable device which can implement any desired circuit by programming the cross-points.

Reconfigurable Computing Hardware

In the last few years there has been a convergence in molecular-scale architectures towards reconfigurable platforms. In the proposed architecture, the nanoFabric, the main computing element is a molecular-based reconfigurable switch. The molecular-based switch eliminates much of the overhead needed to support reconfiguration in traditional CMOS circuits, since the switch holds its own state and can be programmed without extra wires. Researchers exploit the reconfigurable nature of the nanoFabric to provide defect tolerance and to support reconfigurable computing. Reconfigurable computing not only offers the promise of increased performance but it also amortizes the cost of chip manufacturing across many users by allowing circuits to be configured after they are fabricated. Reconfigurable Hardware shares features of both custom hardware and microprocessors. Its computational performance is close to custom hardware, yet, because it is programmable, its flexibility approaches that of a processor.

Researchers are not the first to recognise that RH devices have an enormous potential as computational elements and there has been much research into using RH for computing. Several features of RH devices differentiate them fundamentally from processors, and are responsible for the extreme performance of some computing systems built using such technologies:

1. *Unbounded computational bandwidth:* a microprocessor is designed with a specific number of functional units. The computational bandwidth of a processor is thus bounded at the time of manufacturing. Moreover, it is unusual for a processor to reach its peak performance, because the parallelism available in the programme rarely has the exact same profile as the available functional units. In contrast, RH can support a virtually unbounded number of functional units. Not only can highly parallel computational engines be built, they can exactly fit the application requirements, since the configuration is created post-fabrication.

2. *Unlimited register bandwidth:* another subtle but important difference between a processor and an RH device is in the way they handle intermediate computation results: processors have a predetermined number of registers. If the number of manipulated values exceeds the number of registers, then they have to be spilled into memory. Additionally, the fixed number of internal registers can throttle parallelism.

   Even more limiting is the number of register ports: in a processor the bandwidth in and out of the register file is bounded by a hard limit. In contrast, datapaths on RH directly connect the producers of a result to its consumers. If there is a register needed, it is inserted directly into the datapath. In other words, there is no monolithic register file, no register port limit, and no need to spill values to memory.

3. *Full out-of-order execution:* while superscalar processors allow instructions to execute in orders different from the one indicated by the programme, the opportunity

to do so is actually restricted by several factors, such as limited issue window, generic exception handling and structural hazards. None of these constraints exists in RH implementations. Other often-cited advantages of RH are the abilities to:

— exploit all of an application's parallelism: task-based, data, instruction-level, pipeline, and bit-level,

II. create customised function units and data-paths, matching the application's natural data size,

III eliminate a significant amount of control circuitry.

NanoFabrics

There are two scenarios in which nanoFabrics can be used: (1) as factory-programmable devices configured by the manufacturer to emulate a processor or other computing device, and (2) as reconfigurable computing devices.

(1) In a manufacturer-configured device, user applications treat the device as a fixed processor (or potentially as a small number of different processors). Processor designers will use traditional CAD tools to create designs using standard cell libraries. These designs will then be mapped to a particular chip, taking into account the chip's defects. A finished product is therefore a nanoFabric chip and a ROM containing the configuration for that chip. In this scenario, the configurability of the nanoFabric is used only to accommodate a defect-prone manufacturing process. While this provides the significant benefits of reduced cost and increased densities, it ignores much of the potential in a nanoFabric. Since defect tolerance requires that a nanoFabric be reconfigurable, why not exploit the reconfigurability to build application-specific processors?

(2) Reconfigurable fabrics offer high performance and efficiency because they can implement hardware

> matched to each application. However, this extra performance comes at the cost of significant work by the compiler. A conservative estimate for the number of configurable switches in a 1 cm^2 nanoFabric, including all the overhead for buffers, clock, power, etc., is on the order of 10^{11}. Even assuming that a compiler manipulates only standard cells, the complexity of mapping a circuit design to a nanoFabric will be huge, creating a compilation scalability problem. Traditional approaches to place-and-route in particular will not scale to devices with billions of wires and devices.

A hierarchy of abstract machines that will hide complexity and provide an intellectual lever for compiler designers while preserving the advantages of reconfigurable fabrics. At the highest level is a split-phase abstract machine (SAM), which allows a programme to be broken up into autonomous units. Each unit can be individually placed and routed and then the resulting netlist of pre-placed and routed units can be placed. This hierarchical approach will allow the CAD tools to scale.

Configurable NanoFabric

Using nanoFabrics requires a new manufacturing paradigm, one which trades-off complexity at manufacturing time with post-fabrication programming. The reduction in manufacturing time complexity makes reconfigurable fabrics a particularly attractive architecture for molecular circuits, since directed self-assembly will most easily result in highly regular, homogeneous structures. Effectively utilising the abundant fabric resources and dealing with high defect densities will require a new set of tools. This set includes fast testers to generate defect maps, compilers to convert high-level designs into low-level circuit descriptions, and place-and-route tools to convert circuit descriptions into fabric configurations taking into account the defect map for the fabric. Defects can be tolerated by configuring circuits around them.

A testing step first generates a defect map, much like the maps of defective blocks used for hard disk drives. This map may be a listing of all the defects in the fabric, or may be coarser-grained information such as a count of the number of defects in each portion of the fabric.

Compiler: The compiler takes a high-level design description and compiles it down to a circuit description. Since creating the functionality of the circuit has been moved from manufacturing time to configuration time, the compilation encompasses tasks traditionally assigned to software compilers and to CAD tools.

The challenge of building a single tool spanning both domains presents us with the opportunity of reformulating the traditional interface between these two domains: the Instruction Set Architecture. Removing this rigid interface exposes a wealth of information to the compiler enabling novel classes of optimisations. The disadvantage of this approach is the complexity between the source and target of the compilation process. Some researchers propose the use of traditional, high-level software programming languages for programming nanoFabrics.

The resulting dataflow machines are highly composable: they have simple interfaces using simple, timingindependent protocols. The natural substrate for implementing these dataflow machines are asynchronous (or globally-asynchronous, locally-synchronous) circuits.

Place-and-route: The place-and-route process generates fabric configurations that avoid using defective components. They propose a two-step process to deal with the fabric complexity and the one-of-a-kind nature of each fabric due to defects:

1. An initial, fabric- and defect-independent step (DIPER, or Defect-Unaware Place and Route) which includes technology mapping and some approximate placement and routing, and generates what we call a "soft" configuration. The "soft" configuration is guaranteed to be place-and-routable on a fabric that has a defect level below a certain threshold.
2. A final, defect-aware step (DAPER, or Defect Aware Place and Route) that transforms the "soft" configuration into a "hard" configuration, taking into account the target fabric's defect map. At this step the final place-and-route is performed, using only non-defective resources. (If the defect-map is imprecise, this step requires the use of an on-the-fly defect mapper to pin-point the precise location of defects). The hard configuration is specific to each fabric.

Compilation

We expect future computing systems to be hybrid systems, composed of a general-purpose processor tightly coupled with a reconfigurable fabric. The processor provides excellent virtualisation and will most likely be used for executing the operating system, device interactions and application code with limited parallelism. The RH will be used for compute-intensive kernels, which can take advantage of its wealth of resources.

Application-Specific Hardware (ASH)

There have been many proposals made for implementing hybrid CPU+RH computer systems. We take a fundamentally different approach based on the assumption that the RH will be a significant portion of the total resources.

General: The model supports a general model of computation. It can efficiently execute a wide variety of applications and does not impose unreasonable constraints on the programming model. In fact, the framework should easily accommodate any high-level programming languages available today, e.g. sequential (non-parallel), imperative, pointer-based languages.

Automatic: Programming for an ASH architecture should be no different than programming a processorbased architecture. That is, the compiler should handle all peculiar architectural details, without programmer involvement. Ideally, the programming experience is exactly like traditional programming (and not, for example, like hardware design).

Scalable: ASH is designed to scale with computational resources and clock frequencies. Furthermore, the compiler algorithms are scalable, i.e., they do not rely on excessively complex or expensive whole-programme analysis methods.

Parallel: ASH efficiently exploits the parallelism available in programmes (pipeline, instruction level, data, bit-level, and loop-level parallelism).

Compiling for ASH

Hardware-Software Partitioning: The compiler builds a procedural interface between the code running on the processor and the code implemented directly in hardware. The application programme is partitioned at procedure boundaries; each side invokes the services of the other side using a form of remote procedure call. The partitioner can use multiple sources of information: (1) a conservative estimation of the programme call-graph, (2) the results of a width analysis, which estimates the required hardware resources for implementing each operation, (3) dynamic programme profile information and (4) estimated instruction-level parallelism.

Dataflow Machine Construction: After partitioning, the procedures selected for hardware implementation are processed by the CASH compiler. The output of CASH is a separate hardware circuit for each procedure. CASH essentially transforms the initial C programme into a pure dataflow representation. All C control-flow constructs are transformed into equivalent dataflow operations, as described below. The resulting dataflow representation has semantics very close to that of an asynchronous circuit: data producers handshake with data consumers on data availability and data consumption.

Speculation and Predication: In order to uncover more instruction-level parallelism, and to reduce the impact of control-flow dependences, aggressive predication and speculation are performed on the code. CASH leverages the techniques used in compilers for predicated execution machines: it collects multiple basic blocks into one hyperblock; each hyperblock is transformed into straight-line code by using predication and static single-assignment (SSA). All instructions without side-effects are aggressively predicate-promoted, being speculatively executed. Thus control-flow dependences are reduced to a minimum.

Multiplexors: When multiple definitions of a value reach a join point in the control-flow graph, they must be merged together. This is done by using multiplexors. The multiplexor data inputs are the reaching definitions. The use of decoded multiplexors, which select one ofreaching definitions of a variable; such a multiplexor has 2*n* inputs: one data and one predicate input for each definition. When a predicate evaluates to "true", the multiplexor selects the corresponding input. The hardware implementation of multiplexors and boolean operations is lenient: i.e., as soon as enough inputs have arrived to enable the computation of the output, the output value is immediately generated, without waiting for the remaining inputs. This implementation provides a solution to the problem of unbalanced control-flow created when speculating aggressively.

From Control-Flow to Data-Flow: After hyperblock construction, the only remaining control-flow constructs are inter-hyperblocks transfers, including loops. In other words, each hyperblock is essentially an autonomous unit of SAM. The next compilation phase stitches the hyperblocks together into a dataflow graph representing the whole procedure by creating dataflow edges connecting each hyperblock to its successors. Special "gateway" nodes guide the flow of data from a hyperblock to its successor. This pure-dataflow representation is more detailed than the classical CFG-based representations because the control-flow for each variable is represented separately. This explicit representation naturally exposes loop-level parallelism, and gives rise to loop pipelining at run-time, as described below.

Tokens: The previously described transformations remove all control-flow dependences from the code. The only constraints on the execution order are given by data dependences. However, not all data dependences are explicit in the original programme: operations with side-effects could interfere with each other through memory. In order to maintain programme order for operations whose side-effects do not commute we add edges between instructions that may depend on each other. While many other compilers use such dependence edges internally, in the architecture they are part of the run-time system. Such edges do not carry data values, but they carry an explicit synchronisation token.

The presence of a token at the input is required for these instructions to be executed; on completion, these instructions generate an output token to enable execution of their dependents. A memory operation issues a token to its successors as soon as it has initiated the memory access; it does not need to complete in order to issue the token.

Thus, the tokens are used just to ensure that the memory operations are issued in the right order; dynamic memory disambiguation at the memory side is used to reorder memory accesses accessing different addresses; which naturally supports multiple memories. Token edges explicitly encode may-dependences between memory operations and allow the compiler optimiser to implement very efficiently a wealth of memory-related optimisations, such as redundant operation removal, synchronisation removal and loop pipelining.

Registers: Each operation in the dataflow graph has an associated register on its output. Once the operation has computed its result, it holds it in that register until all consumers are ready to accept it. Only after all consumers acknowledge the receipt of the result, can the operation start a new computation cycle. In this programme representation there is no centralised register file, with multiple read/write ports. Values are stored only at each producer, waiting for the consumer to be ready. As many registers as necessary are synthesised, thus local values are never spilled to memory.

Loop Pipelining: Implementing the data-flow programmes requires no control logic except the local circuitry controlling the handshake between data producers and consumers. The distributed nature of the control logic in a loop provides a natural pipelining effect of the computation. This phenomenon is well known from dataflow machine literature. This effect stems from the fact that the original control-flow-based loop has been decoupled into a separate loop for each loop-carried value. In a processor a single backwards branch instruction signals the start of a new iteration.

The pipeline effect is exploited, for example, by loop induction variables whose values can be quickly computed for several iterations ahead. Long-latency operations within a loop body are then pipelined, receiving in quick succession inputs from the multiple versions of the induction variables. The outcome is that operations from multiple loop iterations proceed simultaneously.

Asynchronous circuits eliminate the global clock, and are based on local communication and synchronisation. The dataflow representation used by CASH is characterised by modularity, distributed control and localised communication. The data-triggered behaviour of dataflow machines is a perfect match for asynchronous circuit implementation.

The use of asynchronous circuits brings other advantages: (1) their modularity and compositionality enables the compiler to scale to large programmes and circuits by using a divide-and-conquer approach for programme translation; (2) the distribution of controlled-skew clock signals to billion-devices circuits in the presence of potential defects in the distribution network is an insurmountable problem; (3) power consumption in asynchronous circuits occurs only in the computationally active parts; (4) placement and routing on synchronous RH must perform timing analysis of all signal paths with respect to the defect map so that the mapped design meets the global clock constraint.

The use of asynchronous designs however brings with it a wealth of research problems. For instance, the contemporary RH devices are all synchronous in nature; such RH platforms and their tool-chains, cannot be directly used to implement asynchronous circuits. Also, the CAD tools for asynchronous design lag far behind the state-of-the-art synchronous design tools.

The radical approach of synthesising asynchronous circuits directly from high-level languages, such as C and FORTRAN. The CASH compiler has a back-end which translates the dataflow intermediate representation into Verilog descriptions of asynchronous circuits. These circuits can then be synthesised and technology mapped for implementation on RH.

The compiler uses a library of modules as the building blocks of the asynchronous circuits. These modules closely correspond to the basic nodes of the intermediate representation, and they encapsulate both data-path and control components.

Every module consists of two parts: the node logic, and control circuitry. The goal of the control circuitry is to achieve flow-control in the absence of the global clock. When input data has arrived and has been processed, this circuit generates a control signal to latch the result in an output register. Flow-control is achieved by establishing a handshaking protocol with neighbours with whom data is exchanged. For this purpose, the bundled data protocol with four-phase handshaking is used.

The bundled data protocol assumes that there exists a pair of wires, request (req) and acknowledgment (ack), associated with each bundle of data wires. The source node, N_s, , sends a signal on the req wire when it has produced data to be consumed by destination, N_d. A property of this protocol is the bundling constraint that requires the data to arrive at N_d before req. After consuming the data, N_d sends a signal on ack. After having received all its inputs and successfully initiated the computation a node can immediately send an ack to its sources.

The handshake naturally makes the circuits latency-tolerant when confronted by either long routes induced by defect tolerant place-and-route or by unpredictable-latency events, such as remote memory access or remote procedure invocations. Modules representing ALU nodes are created by simply synthesising the node logic to perform the corresponding arithmetic or logic function. In order to accommodate simultaneous memory accesses the compiler automatically synthesises a circuit-specific memory channel access arbiter. While traditional processors have one monolithic main memory, with RH we can take advantage of memory locality to create an applicationspecific memory architecture.

The final, synthesised circuit is created by replacing the nodes in the intermediate representation with an instance of the equivalent module from the library. Value-flow edges in the graph are replaced by bundled wire connections. Synthesising these asynchronous circuits differs from prior work in several ways:

1. *Control:* Most work in high-level asynchronous circuit synthesis has concentrated on synthesising asynchronous controllers and state machines. In contrast, the circuits generated by the compiler are dataflow-centric and application-specific, and are composed of arbitrary, non-uniform datapaths. Prior research in asynchronous datapath synthesis has been generally restricted to the design of either specific functional units (adders and multipliers), or the asynchronous processor pipeline.
2. *Toolflow:* In order to reliably build large-scale asynchronous circuits, a tool chain is required that can synthesise such circuits from some high-level language specification. The source language is generally a derivative of Hoare's communicating sequential processes, which describes a set of concurrent processes and their inter-communication. The source language is C, which makes the tool chain accessible to a wider community. The compiler uses powerful analysis techniques to identify parallelism and understand programme behaviour, and then transforms the programme structure to exploit identified opportunities to improve the system performance.

Electronic Nanotechnology

CAEN devices are very small: A single RAM cell will require $100nm^2$ as opposed to $1000nm^2$ for a single laid out CMOS transistor. A simple logic gate or an static memory cell requires several transistors, separate p- and nwells, etc., resulting in a factor of 10^5 difference in density between

CAEN and CMOS. CAEN devices use much less power, since very few electrons are required for switching. CAEN devices are particularly suited for reconfigurable computing since the configuration information for a switch does not need to be stored in a device separately from the switch itself.

On the other hand, A CMOS-based reconfigurable device requires a static RAM cell to controls each pass transistor. Also, two sets of wires are needed in CMOS: one for addressing the configuration bit and one for the actual signals. Perhaps a more realistic comparison of CAEN is to floating-gate technology which also stores the configuration information at the transistor itself.

Like traditional transistors, floating-gate transistors also require two sets of wires. Furthermore, A CAEN switch behaves like a diode, but a floating gate transistor is bi-directional, and thus less useful for building programmable logic. Electronic nanotechnology is quickly progressing and promises incredibly small, dense, and low-power devices. Harnessing this power will require new ways of thinking about the manufacturing process.

Fabrication Process

The process is hierarchical, proceeding from basic components (e.g. wires and switches), through self-assembled arrays of components, to complete systems. In the first step, wires of different types are constructed through chemical self-assembly. The next step aligns groups of wires. Also through self-assembly, two planes of aligned wires will be combined to form a two-dimensional grid with configurable molecular switches at the crosspoints. The resulting grids will be on the order of a few microns.

A separate process will create a silicon-based die using standard lithography. The circuits on this die will provide power, clock lines, an I/O interface, and support logic for the grids of switches. The die will contain "holes" in which the grids are placed, aligned, and connected with the wires on the die. Using only self-assembly and self-alignment restricts us to manufacturing simple, regular structures, e.g., rafts of parallel wires or grids composed of orthogonal rafts. A post-fabrication configuration is used to create useful circuits.

The small size and nondeterministic nature of the self-assembly will also give rise to high defect densities, which can be bypassed through reconfiguration. While researchers have constructed three-terminal EN devices, the precise alignment required to colocate three wires at the device makes them unsuitable for producing real circuits with inexpensive chemical assembly. CAEN devices will be limited to performing logic using two terminal devices; i.e. diode-resistor logic. As the active components will be diodes and configurable switches, there will be no inverters. Because we cannot build inverters, all logic functions will generally compute both the desired output and its complement.

Even more important, the lack of a transistor means that special mechanisms will be required for signal restoration and for building registers. Using CMOS to buffer the signals is unattractive for two reasons: first, CMOS transistors are significantly larger and would decrease the density of the fabric. Second, the large size of CMOS transistors would slow down the nanoFabric. The fabrication process also disallows the precise alignment required to make end-to-end connections between nanoscale wires.

Architecture of NanoFabric

The architecture of the nanoFabric is designed to overcome the constraints associated with directed assembly of nanometer-scale components and exploit the advantages of molecular electronics. Because the fabrication process is fundamentally non-deterministic and prone to introducing defects, the nanoFabric must be reconfigurable and amenable to self-testing. This allows us to discover the characteristics of each nanoFabric and then create circuits that avoid the defects and use the available resources.

For the foreseeable future, the fabrication processes will only produce simple, regular geometries. Therefore, the proposed nanoFabric is built out of simple, two-dimensional, homogeneous structures. Rather than fabricating complex circuits, the researchers use the reconfigurability of the fabric to implement arbitrary functions post-fabrication. The construction process is also parallel; heterogeneity is introduced only at a lithographic scale. The nanoFabric can be configured (and reconfigured) to implement any circuit, like today's FPGAs; the nanoFabric though has several orders of magnitude more resources.

The nanoFabric is a 2-D mesh of interconnected nanoBlocks. The nanoBlocks are logic blocks that can be programmed to implement a three-bit input to three-bit output Boolean function and its complement. NanoBlocks can also be used as switches to route signals.

The nanoBlocks are organised into clusters. Within a cluster the nanoBlocks are connected to their nearest four neighbors. Long wires, which may span many clusters (long-lines), are used to route signals between clusters. The nanoBlocks on the perimeter of the cluster are connected to the long-lines. This arrangement is similar to commercial FPGAs (allowing us to leverage current FPGA tools) and has been shown to be flexible enough to implement any circuit on the underlying fabric.

The nanoBlock design is dictated by fabrication constraints. Each side of the block can have inputs or outputs, but not both. We have arranged it so that all nanoscale wire-towire connections are made between two orthogonal wires; we do not need precise end-to-end alignment. Notice that the outputs of the blocks are either facing south and east (SE) or north and west (NW). By arranging the blocks such that all the SE blocks run in one diagonal and the NW run in the adjacent diagonal we can map any circuit netlist onto the nanoFabric. Since the nanoBlocks themselves are larger than the minimum lithographic dimension (e.g, greater than one micron), they can be positioned precisely at manufacturing time in the desired patterns.

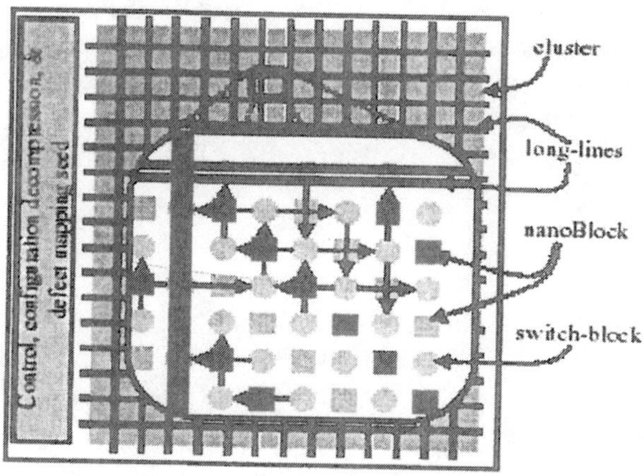

Figure 1. The layout of the nanoFabric with a partial blowup of a single cluster and some of the adjacent long-lines.

In addition to the intra-cluster routing there are longlines that run between the clusters to provide low-latency communication over longer distances. The nanowires in these tracks will be of varying lengths, allowing a signal to traverse one or more clusters without going through any switches. This layout is essentially that of an island style FPGA. This general layout has been shown to be efficient and amenable to place-and-route tools. Notice that all communication between nanoBlocks occurs at the nanoscale. The fact that we never need to go between nanoscale and CMOS components and back again increases the density of the nanoFabric and lowers its power requirements.

The arrangement of the clusters and the long-lines promote scalability in several ways. First, as the number of components increases we can increase the number of longlines that run between the clusters. This supports routability of netlists. Second, each cluster is designed to be configured in parallel, allowing configuration times to remain reasonable even for very large fabrics. The power requirements remain low because we use molecular devices for all aspects of circuit operation. Finally, because we assemble the nanoFabric hierarchically we can exploit the parallel nature of chemical assembly.

NanoBlock

The nanoBlock is the fundamental unit of the nanoFabric. It is composed of three sections (Figure 2): (1) the molecular logic array, where the functionality of the block is located, (2) the latches, used for signal restoration and signal latching for sequential circuit implementation, and (3) the I/O area, used to connect the nanoBlock to its neighbors through the switch block.

The molecular logic array (MLA) portion of a nanoBlock is composed of two orthogonal sets of wires. At each intersection of two wires lies a configurable molecular switch. The switches, when configured to be "on", act as diodes. Designing circuits for the MLA is significantly different than for a programmable logic array, which requires an OR and an AND plane. We have preliminary designs for a "standard cell" library using nanoBlocks, e.g. AND, OR, XOR, and ADDER. Figure 3 shows the implementation of an and gate.

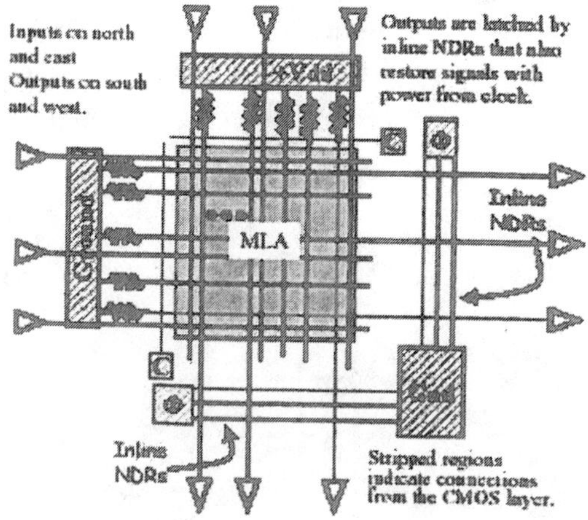

Figure 2. A schematic of a nanoBlock.

Notice that all the connections between the CMOS layer and the nanoBlock occur either between groups of wires or with a wire that is removed from all the other components. This improves the device density of the fabric. To achieve a specific functionality the cross-points are configured to be either open-connections or to be diodes. The layout of the MLA and of the switch block makes rerouting easy in the presence of faults.

In fact, the rows can be moved anywhere within the block without affecting the circuit, which makes defect tolerance signifi-cantly easier than with CMOS. The number of possible ways to arrange the columns/rows in the MLA combined with the configurable crossbar implemented by the switch block makes the entire design robust to defects in either the switch block or the MLA.

The nanoFabric is defect tolerant because it is regular, highly configurable, fine-grained, and has a rich interconnect. The regularity allows us to choose where a particular function is implemented. The configurability allows us to pick which nanowires, nanoBlocks, or parts of a nanoBlock will implement a particular circuit. The fine-grained nature of the device combined with the local nature of the interconnect reduces the impact of a defect to only a small portion of the fabric (or even a small portion of a single nanoBlock).

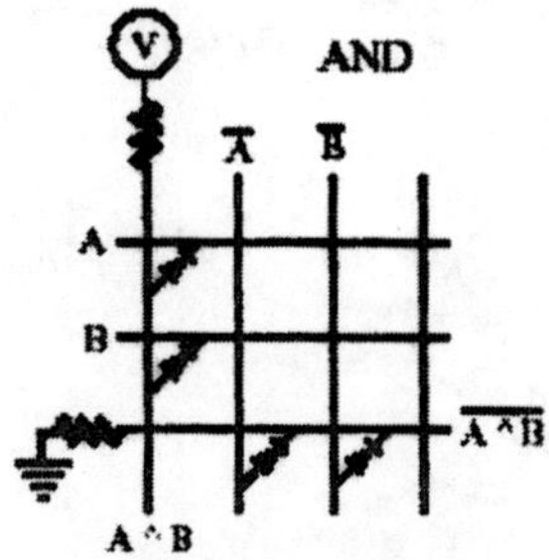

Figure 3. An AND gate implemented in the MLA of a nanoBlock.

Finally, the rich interconnect allows us to choose among many paths in implementing a circuit. Thus, with a defect map we can create working circuits on a defective fabric. Defect discovery relies on the fact that we can configure the nanoFabric to implement any circuit, which implies that we can configure the nanoFabric to test its own resources. The key difficulty in testing the nanoFabric (or any FPGA) is that it is not possible to test the individual components in isolation. Researchers on the Teramac project faced similar issues. They devised an algorithm that allowed the Teramac, in conjunction with an outside host, to test itself. Despite the fact that over 75% of the chips in the Teramac have defects, the Teramac is still used today.

The basis of the defect mapping algorithm is to configure a set of devices to act as tester circuits. These circuits, e.g. linear-feedback shift-registers, will report a result which if correct indicates with high probability that devices they are made from are fault free. The defect mapping process is complicated by the fact there are no known fault-free regions in the nanoFabric at the start of testing and as the defect density increases the chances of finding a good circuit decreases. To address this problem, we will implement in CMOS the components of a basic tester, which will provide the initial fault-free region. A host computer configures testers out of the CMOS testor and portions of the nanoFabric. As the host gains knowledge of the fault free regions of the nanoFabric it replaces more and more of the CMOS tester with configured sections of the nanoFabric. After a sufficient number of functioning resources have been discovered by the host, phase two of the testing begins.

For this phase, the host downloads a configuration, which performs the testing from within the fabric. In other words, the already tested area of the fabric acts as a host for testing the remainder of the fabric. Because the number of tests required to isolate any specific defect does not grow as the total size of the device grows, the computational work needed to test a device is at worst linear in the size of the device. For very large devices, such defect discovery can be accomplished using many parallel independent test machines. Once a defect map has been generated the fabric can be used to implement arbitrary circuits.

The architecture of the nanoBlock supports full utilisation of the device even in the presence of a significant number of defects. Due to the way we map logic to wires and switches, only about 20% of the switches will be in use at any one time. Since the internal lines in a nanoBlock are completely interchangeable, we generally should be able to arrange the switches that need to be configured in the ON state to be on wires which avoid the defects. While the molecules are expected to be robust over time, inevitably new defects will occur over time. Finding these defects, however, will be significantly easier than doing the original defect mapping because the unknown defect density will be very low.

The nanoFabric uses runtime reconfiguration for defect testing and to perform its intended function. Thus, it is essential that the time to configure the fabric scale with the size of the device. There are two factors that contribute to the configuration time. The first factor is the time that it takes to download a configuration to the nanoFabric.

The second factor is the time that it takes to distribute the configuration bits to the different regions of the nanoFabric. Configuration decoders are required to serialise the configuration process in each nanoBlock. To reduce the CMOS overhead, initially we intend to only configure one nanoBlock per cluster at a time. However, the fabric has been designed so that the clusters can be programmed in parallel. A very conservative estimate is that we can simultaneously configure one nanoBlock in each of 1000 clusters in parallel.

A molecular switch is configured when the voltage across the device is increased outside the normal operating range. Devices in the switch blocks can be configured directly by applying a voltage difference between the long intercluster lines. In order to achieve the densities presented above, it will also be necessary to develop a configuration approach for the switches in the MLA that is implemented with nanoscale components. In particular, a nanoscale decoder is required to address each intersection of the MLA independently. Instead of addressing accessing each nanoscale wire separately in space we address them separately in the time dimension.

This slows down the configuration time, but increases the device density. Preliminary calculations indicate that we can load the full nanoFabric, which is comprised of 10^9 configuration bits at a density of 10^{10} configuration bits/cm^2, in less than one second. This calculation is based on realistic assumptions that, on average, fewer than 10% of the bits are set ON and that the configurations are highly compressible. It also significant to note that it is not necessary to configure the full fabric for defect testing. Instead, we will configure only the portions under test.

As the configuration is loaded onto the fabric it will be used to configure the nanoBlocks. Using the configuration decoder this will require ≈300 cycles per nanoBlock, or less than 38K cycles per cluster. Thus, the worst-case time to configure all the clusters at a very conservative 10 MHz requires three seconds.

The nanoFabric is a reconfigurable architecture built out of CMOS and CAEN technology. The support system, i.e., power, ground, clock, and configuration wires, I/O, and basic control, is implemented in CMOS. On top of the CMOS the researchers construct the nanoBlocks and long-lines constructed out of chemically self-assembled nanoscale components.

Assuming a 100nm CMOS process and 40nm centers with 128 blocks to a cluster and 30 long-lines per channel, the design should yield 750K clusters/cm^2 (or 1M blocks/ cm^2), requiring 3^{10} configuration bits. If the nanoscale wires are on 10nm centers this design yields 1M clusters/cm^2. SPICE simulations show that a nanoBlock configured to act as a half-adder can operate at between 100MHz and 1GHz. Preliminary calculations show that the fabric as a whole will have a static power dissipation of ≈ 1.2 watts and dynamic power consumption of ≈.4 watts at 100Mhz.There are two scenarios in which nanoFabrics can be used: as factory-programmable devices configured by the manufacturer to emulate a processor or other computing device, and as reconfigurable computing devices. In a manufacturer-configured device, user applications treat the device as a fixed processor.

Processor designers will use traditional CAD tools to create designs using standard cell libraries. These designs will then be mapped to a particular chip, taking into account the chip's defects. A finished product is thus a nanoFabric chip and a ROM containing the configuration for that chip. In this mode, the configurability of the nanoFabric is used only to accommodate a defectprone manufacturing process. While this provides the significant benefits of reduced cost and increased densities, it ignores much of the potential in a nanoFabric. Since defect tolerance requires that a nanoFabric be reconfigurable why not exploit the reconfigurability to build applicationspecific processors? Reconfigurable fabrics offer high performance and efficiency because they can implement hardware matched to each application.

Further, the configurations are created at compile time, eliminating the need for complex control circuitry. Research has already shown that the ability of the compiler to examine the entire application gives a reconfigurable device efficiency advantages because it can:

- exploit all of an application's parallelism: MIMD, SIMD, instruction-level, pipeline, and bit-level.
- create customised function units.
- eliminate a significant amount of control circuitry.
- reduce memory bandwidth requirements.
- size function units to the application's natural word size.
- use partial evaluation and constant propagation to reduce the complexity of operations.

However, this extra performance comes at the cost of significant work by the compiler. A conservative estimate for the number of configurable switches in a 1cm^2 nanoFabric, including all the overhead for buffers, clock, power, etc. is on the order of 10. Even assuming that a compiler manipulates only standard cells, the complexity of mapping a circuit design to a nanoFabric will be huge, and this creates a compilation scalability problem. Traditional approaches to place-and-route in particular will not scale to devices with billions of wires and devices.

Split-Phase Abstract Machine (SAM)

The compilation process starts by partitioning the application into a collection of threads. Each thread is a sequence of instructions ending in a split-phase operation. An operation is deemed to be a split-phase operation if it has an unpredictable latency. For example, memory references and procedure calls are all split-phase operations. Thus, each thread, similar in spirit to a Threaded Abstract Machine (TAM) thread, communicates with other threads asynchronously using split-phase operations. This partitioning allows the CAD tools to concentrate on mapping small isolated netlists and it has all the mechanisms required to support thread-based parallelism. Unlike a traditional thread model, where a thread is associated with a processor when executing, each SAM thread will be a custom "processor." While it is possible for a thread to be complex and load "instructions" from its local store, the intention is that it remains fairly simple, implementing only a small piece of a procedure.

This allows the threads to act either in parallel or as a series of sequential processes. It also reduces the number of timing constraints on the system, which is vital for increasing defect tolerance and decreasing compiler complexity. The SAM model is a simplification of TAM. A SAM thread/processor is similar to a single-threaded codeblock in TAM, and memory operations in SAM are similar to memory operations in Split-C. In a sense, SAM will implement Active Messages for all interprocessor communications. While SAM can support parallel computation, a parallelising compiler is not necessary. The performance of this model rests on the ability to create custom processors. A compiler could construct machines in which only one processor is active at a time.

Later, as the compiler technology becomes more mature, the inherently parallel nature of the model can be exploited. The SAM model explicitly hides many important details. For example, it neither addresses dynamic routing of messages nor allocation of stacks to the threads. Once an application has been turned into a set of cooperating SAM threads it is mapped to a more concrete architectural model which takes these issues into account.

Nanomemory Systems

Within only the last few years, the field of molecular electronics has seen the demonstration of singlemolecule electrical devices and the subsequent construction of functioning simple molecular-scale circuits, including a few small nanomemory arrays. Additional attempts to produce extended circuit systems are underway. As one such example, the U.S. Government's Defense Advanced Research Projects Agency (DARPA) is attempting to develop and build ultradense computer memories and processors from molecular-scale components.

More specifically, DARPA's Moletronics Programme is intent on delivering a 16-kilobit nanomemory with a density of 10^{11} bits/cm^2 by the year 2005. By comparison, the International Technology Roadmap for Semiconductors (ITRS) estimates commercial DRAM densities will just exceed 10^9 bits/cm^2 in 2004 and be slightly less than 3×10^{10} bits/cm^2 by 2016. Despite these aggressive goals, it is still an open question whether such a densely integrated, extended memory system can be made to operate if it is built using existing nanodevices and nanofabrication technologies.

Simulations and analysis of the experimental nanomemory being developed by Harvard University and the California Institute of Technology (Caltech) under the DARPA Moletronics Programme. These simulations suggest that the planned 16-Kilobit nanomemory array will operate effectively. The simulation methodology consists of constructing device models of experimental molecular-scale electronic devices, for example, diodes and transistors, that already have been demonstrated at Harvard University. These novel device models are then incorporated into conventional very large-scale integration computer aided design (VLSI CAD) simulations of the complete nanomemory system.

The primary focus of this simulation research has been to explore the impact on nanomemory function as the bit count is scaled from that of presently fabricated prototypes, with only a few bits, to the 16-Kbit array size required by the Moletronics Programme. In addition, a variety of other design parameters, for example, capacitance to ground and banking topology, have also been investigated in order to provide information to the developers at Harvard and Caltech in advance of actual experiments.

The Harvard-Caltech nanomemory system design consists of three major subsystems: a nanowire crossbar memory array and two decoders, one for the array rows and one for the columns. Each nanowire-nanowire crosspoint in the crossbar array forms a bistable nonvolatile nanowire (NVNW) diode that can store one bit of information each.

The microwire-nanowire crosspoints in the decoders form field-effect transistors, similar to the crossed nanowire transistors described by Huang et al. These transistors permit the selection of individual rows and columns in the memory array. Addressing the nanomemory array is accomplished with a "2-hot" encoding scheme. This encoding scheme differs from the binary schemes typically used in complementary metal-oxide-semiconductor (CMOS) circuitry in that the 2-hot scheme requires asserting exactly two microwires to select a unique nanomemory address, no matter the size of the address space.

The functionality is defined as the ability to correctly read and write information. To evaluate read operations, on the output current differences (ΔI_{out}) between reading logic "1" and reading logic "0" for the worst-case memory configurations. For write operations, to determine that the selected diode, and only the selected diode, is subjected to the appropriate voltage bias for storing the logic value. Another important design goal of the nanomemory is ultrahigh density.

The DARPA Moletronics Programme goal of a 16-kilobit memory with a density of 10^{11} bits/cm^2 means that the nanomemory can occupy an area of only 16μm^2. Therefore, analysis of the nanomemory footprint is warranted. Simple area equations are used to evaluate these parameters. Finally, consideration is given to the nanomemory speed and power consumption. Output current switching times are analysed, and methods are suggested to improve speed and reduce power. However, a thorough treatment of these issues requires additional information about the peripheral circuitry surrounding the nanomemory.

The simulation of devices and complex circuit systems can be performed at a number of different levels of design abstraction. The appropriate design abstraction depends largely on the goals of the simulation, capturing the important behaviours of the overall system, while disregarding immaterial details.

There are three general categories of design abstraction: the device level, the circuit level, and the architectural level. The device level focuses on a single device, for example, a diode or transistor, in great detail. Such simulations provide information about the operation and physics of individual devices, but generally do not consider the interactions among devices in a circuit. At the other end of the spectrum, the architectural level considers very large systems, but typically does not include the physics or the behaviour of individual devices.

In contrast, the circuit level bridges these two approaches and considers relatively large systems (on the order of tens of thousands of devices), while still retaining a connection to the underlying physical behaviour. The simulations described here take place at this level. Many concepts and techniques in conventional microelectronics are borrowed to simulate the nanoelectronic memory. For example, the commonly used commercial Cadence Spectre VLSI CAD software tool is the primary simulation programme.

One reason for applying such commercial off-the-shelf software tools from the microelectronics industry is the obvious timesaving and reliability associated with the use of readily available, well-tested software. This software also incorporates powerful features, such as modeling languages and graphics, developed specifically for the flexible modeling of extended circuitry.

Finally, using conventional VLSI tools provides a seamless approach to the design and simulation of both the nanomemory and the peripheral microelectronic circuitry that it requires for operation and communication with the outside world. It relies heavily on the conventional microelectronics concept of the device model, which captures the essential properties and response behaviour of a circuit element.

Models of the experimentally observed behaviour of all of the devices contained in the nanomemory are required. In particular, the current- voltage transfer characteristics (I-V curves) are necessary for steady state (DC) simulation, and the capacitance-voltage transfer characteristics (C-V curves) are required for time-varying, or transient, simulation. Typical device models for microelectronics consist of compact equations based upon the underlying physics of a device that has been extensively studied and is well understood. However, this physics based approach is not workable, at present, for simulations involving molecular-scale devices, because the fundamental physics of most molecular-scale devices is not well known.

Incorporating new models into conventional circuit simulators can be difficult. The addition of a new model can often require modifying proprietary source code. Open source simulators do exist, for example, SPICE3, but adding new device models is tedious. Furthermore, these open-source simulators lack the robustness and simulation speed that are required to model large circuit systems, and which are found in many commercial simulators. Thus, to develop and simulate models for molecular-scale devices efficiently, to use a commercial simulator and to fit or describe the empirical behaviour of the devices using the analog hardware description language (analog HDL) Verilog-A. This modeling approach is similar to one described elsewhere, except that the empirical equations derived here are tailored to the devices employed in the Harvard-Caltech nanomemory system.

The empirical equations were incorporated into the Spectre circuit simulator from Cadence, which supports cosimulation of both Verilog-A components and conventional SPICE-level devices. The combination of the empirical device modeling with Verilog-A, and the ease of incorporation of these models into the Spectre simulator, made it possible to model quickly novel molecular-scale devices, simulate circuits composed of these new models, and incorporate state-of-the-art conventional microelectronic device models, as necessary. The simulations of the Harvard-Caltech crossbar nanomemory required three device models. The first two are models of the distinct nanowire devices described in the previous section: the crossed nanowires that form the nonvolatile nanowire (NVNW) diodes and the top-gated nanowire FETs (TGNW-FETs). The third is a model of the nanowire interconnects of the nanomemory.

Nanowire Interconnect Model

In conventional microelectronics, there is a clear-cut distinction between the devices and the wires that connect them. This distinction does not exist in the crossbar nanomemory considered here. Nanowires in the nanomemory form the devices, as well as connect the devices to one another. To deal with these two roles, an interconnect model was developed that incorporates an artificial division between the two functions. An important resistance in the interconnect model that is not shown in the figure is the contact resistance R_c between the microwire power supplies and the nanowires. This value is approximately 1M. in the present devices. Furthermore, R_c is dominant in comparison to RNW. Thus, the nanowire resistance is negligible in the simulations.

Although the parameter *RNW* is not employed in the simulations presented, including it in the interconnect model provides the capability to account for the nanowire resistance when improvements in the fabrication techniques reduce R_c to a value where the two resistances are comparable. The capacitance to the substrate CN_{Wsub} is treated as a variable parameter.

The actual experimental value can be altered by the changing the nanomemory distance from the substrate and/ or by changing the insulating dielectric between the crossbar and substrate. This value has an important influence on the system performance. There are two additional parasitic influences that may play a role in nanoscale systems and that were not included in the nanowire interconnect model.

Neither crosstalk capacitance between neighbouring wires nor parasitic inductances along the wires have been incorporated into the model, thus far. These two effects may manifest themselves in systems with small wire pitches or in systems with long and narrow wires operating at high frequencies, respectively. However, these effects will not influence strongly the functionality of a low speed, low frequency prototype nanomemory. In other words, whereas these two parasitics may impact the operating conditions and efficiency of the nanomemory, they will not alter whether or not the system can be made to operate. Nevertheless, since these effects may influence the optimisation of the nanomemory, simulations that explicitly incorporate these effects are being developed and will be included in future research efforts.

Banking Topologies

The simply increasing the size of a single nanomemory array may not be the most effective approach for producing memories with very high bit counts. As the size of a memory arrays increases, so do the resistances and capacitances associated with the array, which increase delay and power consumption. Ultimately, this may threaten functionality. In addition, large memory arrays are more susceptible to fabrication defects, since a single defect in a wire can render all the memory cells along it unusable. Reducing the vulnerability of nanomemories to defects is important. This is because, based on statistical and thermodynamic arguments, it is anticipated that the hierarchical self-assembly strategies being pursued for molecular-scale electronic circuits may produce a significant fraction of defective devices or devices that are imprecisely positioned. To increase defect and fault tolerance, instead of using a single large array to achieve a high bit count, banks of smaller memories might be employed.

Adopting a banking strategy also increases the overall data throughput for the memory. First, the lower resistances and capacitances of the shorter nanowires in the smaller arrays allow faster access times. Second, banked arrays can be accessed in parallel (i.e., a bit can be accessed from each bank simultaneously) significantly increasing memory performance. Although banked architectures can create more complex fabrication patterns, the regularity of the banks would seem to provide a feasible route to nanomemory assembly. Harvard University already has made significant progress in the parallel fabrication of multiple arrays in a tiled pattern. The one significant trade-off generally associated with employing a banking strategy is an increase in area per usable bit. This occurs because each additional bank requires additional wires for encoding and decoding the memory array.

Although some of these wires can be shared among the banks, banking always results in an increase in the number of address wires. Thus, an optimal banking strategy will employ moderately sized arrays that take advantage of the coding scheme to increase density, but also achieve the requisite degree of defect tolerance, parallel access, and/or other design goals. Despite the various possible banking topologies for producing a 16-kilobit nanomemory, first-order area calculations suggest that the target nanowire pitch should be similar for a variety of topologies.

7

Economic Impact of Nanotechnology

In examining nanoscale objects or in making specific structural changes, scientists generally use physical methods. The production of nanoscale particles, on the other hand, occurs primarily within the realm of chemistry. Nano-objects such as structural proteins, enzymes and viruses, however, are created by self-organisation according to the laws of nature, and a large proportion of the basic processes, such as cellular energy generation processes (such as the respiratory chain or photosynthesis), occur on the nanoscale, i.e. at the molecular level. Bridging the gap between the macro, micro and nanoworlds is the task of system integration techniques.

The required systems technologies include design and simulation tools and methods, assembly and joining technology, test methods, and appropriate production processes. Important demands on the assembly and joining technologies at the micro-nano interface include the realisation of the necessary mechanical, electromechanical or biochemical connections as well as the use of microsystems technology in the positioning and wiring of nano-components.

As a result, a boom of innovations is expected in virtually all high-technology industries, for example in information and communications technology, in automotive, power and production engineering, in the chemical and pharmaceutical industries and in medicine and biotechnology. Society will then be able to enjoy the prospects of increased environmental compatibility of lifestyle and mobility, drastically improved communications and information as well as optimised healthcare.

Nanotechnology can contribute to better quality of care at less cost in an aging society through improved and more economical diagnostic and treatment methods, including nanostructured biochips, nanoprobes, intelligent depot medications for sustained release, microsystems to complement organic functions, and artificial substrate materials for tissue implants. Even today it is possible to envision examples of future applications in these areas that are likely to benefit substantially from the advances in nanotechnology.

Automotive engineering is pursuing applications to which nanotechnology can make a significant contribution. In this quest, new technical refinements are useful both functionally and cosmetically. The car of the future will respond intelligently both to environmental stimuli and to driver behaviour. Windows and mirrors will adjust to exterior lighting. Tyres will have good traction on the most diverse road surfaces. And multiple sensors will proactively adjust the driving condition to a change in weather or an impending collision. Switchable colour changes in the paint and easier modifications using lightweight designs will allow customised styling.

New knowledge in nanotechnology will contribute to optimised combustion and emission control, a reduction in body weight, the development of self-healing paint, wear-resistant tyres with good road grip, and functional window glass. Electronics already contribute a disproportionately large share to the added value in automobile manufacturing. The importance of automotive electronics will continue to grow and, in keeping with the auto industry's role as a technology driver, spur on the development of nanoelectronic innovations into marketable products.

Machine and plant construction contributes to the advancement of nanotechnology applications in two different contexts. On the one hand, this discipline makes new manufacturing and systems technologies available. On the other, the use of nanostructures in function-determining layers, in measuring instruments and in sensing systems makes it possible to design better machines and plants. As a result, the productivity and reliability of machines and plants is increased, making this traditionally strong German industry and its products even more competitive.

The development of highly efficient or even autonomous power supplies for portable devices is an urgent priority in power engineering. Disposable as well as rechargeable batteries continue to be unsatisfactory with respect to weight and performance. But a combination of economical electrical and electronic components, efficient lighting and display systems, and above all innovative energy storage devices such as small-format fuel cells is expected to provide new solutions. What's more, integrated energy converters based on nanotechnology, such as highly efficient solar cells or innovative materials that directly convert heat into electricity may provide a replacement for rechargeable batteries in low-power applications. That would give an enormous boost to products such as portable electronics and diagnostic systems.

In medicine, the focus continues to be on diagnostic and therapeutic approaches for managing widespread diseases such as cancer and diabetes. In this field, nanoparticles represent a new platform technology. They can be accumulated in the tumour as a selective contrast medium for medical imaging, or destroy tumour tissues locally by heating them, or, even more importantly, by transporting specifically effective therapeutic agents.

In the long term, these particles also appear to point the way to noninvasive methods for early diagnosis. What's more, these techniques will be complemented by diagnostic applications of biochips produced by nano and microsystems technology. A virtual boom is already under way in the use of these chips in pharmaceutical research and laboratory applications. The expansion of biochip technologies brings us a significant step closer to the distant goal of individualised medicine, in which prompt on-site diagnostics will be complemented by customised medications. The first drug delivery systems for transporting chemotherapeutic agents to the tumour are now close to being approved.

In the field of information and communications, the availability of any desired information, any time, anywhere in the world, would be an objective toward which nanotechnologies, and especially nanoelectronics, can make a significant contribution. In the future one should be able to use a device of negligible size to access any required information, or to communicate with any desired party. In such devices, the information should not merely appear as a string of characters on a tiny display. Instead, the user should be able to select the preferred presentation from a diversity of processed formats, ranging perhaps from text-to-speech all the way to three-dimensional holographic images. These devices would provide medical on-line diagnostic capability with automatic alerts, and their data storage would have the capacity of a national library. The computing power of such an "electronic assistant" would rival that of a present-day computing centre.

In optics too, developments are opening up a wide range of potential future products. Optoelectronic components already play an important part in home entertainment (CD/DVD, laser TV, projection systems, optical interconnects) and will continue to grow in importance, as will lighting technologies based on optoelectronic components, such as large area light-emitting diodes (LEDs). Due to their long service life and high efficiency, optoelectronic light sources are not only more reliable than conventional light sources; they also consume much less energy and are therefore more cost-efficient. What's more, optical lenses of any desired geometry and manufactured with nanometre precision will be used for highly precise, function-specific conduction and deflection of light in applications such as data equipment and (home) cinema projectors, lithography or medical systems.

Global Networking Activities

During the course of the past decade, advances in our understanding of quantum effects, boundary properties and surface properties, as well as of the principles of self-organisation, have laid the foundations for innovative analytical and production techniques which have caused an upsurge of interest in nanotechnology and global networking activities along the value-added chain. This interest has been boosted in particular by the early combination of basic research results and application options, and by the resulting expectations regarding potential markets.

The players on Germany's nanotechnology scene were among the world's first to address potential applications at an early stage, on the basis of solid and broad-based fundamental research. More than 100 companies in Germany have already recognised these innovation opportunities and are using nanotechnology know-how in their core business. Today, a total of about 400 to 500 companies in Germany are involved with nanotechnology and are becoming increasingly active in this field as product developers, suppliers or investors. These companies do not view nanotechnological R&D work as a short-lived fashion but are taking a long-term approach in addressing key elements for future innovation in industries with a large job-creation potential, primarily in the automotive and machine-construction industries, in chemicals and pharmaceuticals, in the optical industry, medicine and biotechnology, as well as in power generation and construction.

Many small and medium-sized enterprises (SMEs) that can be ranked as pure nano businesses have sprung up in Germany. These flexible innovation companies occupy specific niches in the value-added chain and make an important contribution to know-how transfer from research to industry. SMEs consequently serve a key function in most high-technology fields, and establishing innovative start-ups is therefore of enormous importance in the young nanotechnology industry too. Since the late 1980s, the BMBF has been funding nanotechnology research activities in the contexts of its Materials Research and Physical Technologies programmes.

Initial core topic areas included the production of nanopowders, the creation of lateral structures on silicon and the development of nanoanalytical methods. BMBF support was later expanded to also include other programmes with relevance to nanotechnology for instance in the Laser Research and Optoelectronics programmes. Today, many projects related to nanotechnology are supported through a considerable number of specialised programmes. Examples include Materials Innovations for Industry and Society (WING), IT Research 2006, the Optical Technologies Sponsorship Programme and the Biotechnology Framework Programme. From 1998 to 2004, the volume of funded joint projects in nanotechnology has quadrupled to about €120 million.

In addition to BMBF-funded research, project-related investments are also financed by the Ministry of Economics and Employment (BMWA) in the Physikalisch-Technische Bundesanstalt (PTB—the national metrology institute) and the Federal Institute for Materials Research and Testing

(BAM), as well as nanotechnology-related projects in the PRO INNO innovation competency programme for SMEs. These projects are funded to the tune of about ¢ 25 million annually.

The purpose of the infrastructural activity of these competence centres is to optimise the conditions for bringing potential users and nanotechnology researchers together. The centres will efficiently focus the nanotechnological knowledge of their members and convert it into industrial development.

Other tasks of the competence centres include in particular activities related to training and continuing education, collaboration on issues concerning standardisation and regulations, consulting and support of would-be entrepreneurs and public-relations work. The individual competence centres are organised along the subject-specific value-added chain in their respective fields.

The entire network presently interlinks approximately 440 players from the academic sphere (29%), research institutions (23%), large corporations (12%), small and medium-sized enterprises (31%) as well as financial services, consultants and associations (totalling 5%). The information sharing supported by the competence centres is particularly helpful for small companies to keep them informed about current developments and what such developments mean to them. In the next three years, the centres intend to focus especially on training and continuing education, and on supporting start-up companies.

The competence centres that are directly supported by the BMBF, several other networks have evolved that pursue different goals and are therefore differently structured. In contrast to networks with a structure that is generally nationwide, several universities and research centres have consolidated their nanotechnological basic research activities through local—in some cases even internal—networks.

In Germany, institutional research in nanotechnology outside the universities is pursued by the four large research associations: MPG, FhG, HGF and WGL. These associations maintain numerous research establishments or working groups whose range of activities includes nanotechnology research.

Wissenschaftsgemeinschaft G. W. Leibniz (WGL): The institutes of the WGL (G.W. Leibniz Science Association) conduct basic and industrially orientated work in nanotechnology. Some of their principal efforts are focused on nanomaterials research, in which the Institutes for New Materials (INM , Saarbrücken), for Solid State and Materials Research (IFW, Dresden) and for Polymer Research (IPF, Dresden) rank among the leaders; and on surface technology, for instance at the Institute for Surface Modification (IOM, Leipzig) and at the Rossendorf Research Centre (FZR). Work at the Paul-Drude-Institut (PDI, Berlin) includes basic research in solid state electronics.

Helmholtz-Gemeinschaft deutscher Forschungszentren (HGF): The HGF (Helmholtz Association of National Research Centres) also conducts work on issues related to materials and nanoelectronics. Especially noteworthy is the work at the two research centres in Karlsruhe (FZK) and Jülich (FZJ). R&D on nanomaterials and thin-film systems is also pursued at the research centre in Geesthacht (GKSS) and at the Hahn-Meitner-Institut in Berlin (HMI).

Max-Planck-Gesellschaft (MPG): Work at the institutes of the MPG (Max Planck Society) is contributing fundamentally important knowledge towards new approaches in nanotechnology research. The Institute for Solid State Research and Metals Research in Stuttgart and the MPI for Microstructure Physics in Halle for instances have been active for many years in the fields of nanomaterials, characterisation methods and new functionalities.

Internationally recognised R&D achievements have also been contributed by the Institutes for Polymer Research (Mainz), for Colloid and Boundary Layer Research (Golm), for Biochemistry (Munich-Martinsried), for Coal Research (Mülheim), and by the Fritz-Haber Institute (Berlin).

Fraunhofer Gesellschaft (FhG): Since an industrial demand already exists in most areas of nanotechnology, many institutes of the FhG (Fraunhofer Society) conduct projects focused on specific applications jointly with industrial companies. Some of these efforts are focused on thin-film and surface technologies, a field in which the FhG has been supported by the BMBF for many years.

The Institutes for Materials and Laser Beam Technology (IWS, Dresden), for Silicate Research (ISC, Würzburg), for Optics and Precision Mechanics (IOF, Jena) and for Boundary Layer Research (IGB, Stuttgart) have been very active in this subject area. Nanomaterials research receives high priority at the Institutes for Applied Materials Research (IFAM, Bremen), for Applied Solid State Physics (IAF, Freiburg) and for Chemical Technology (ICT, Pfinztal), among others. The Institutes for Silicon Technology (ISIT, Itzehoe) and for Production Technology (IPT, Aachen) are exploring the interface of microtechnology and nanotechnology.

The Fraunhofer Institute for Reliability and Microintegration (IZM, Berlin) is making contributions in particular to assembly and interconnection technology. The Institute for Biomedical Technology (IBMT, St. Ingbert) is exploring links to nanobiotechnology. The Institute for Solar Energies (ISE, Freiburg) is investigating the contribution of nanotechnology to energy production.

Nearly all German universities with a technical and scientific programme of studies are conducting R&D related to nanotechnology. At the same time, growing emphasis is given to developing an interdisciplinary understanding of the relationships in various areas of this field. At several universities, nanotechnology courses of study have already been established that are closely linked to current research topics. Examples of the comprehensive activities covering this subject area can be found in the academic centres of Karlsruhe, Aachen, München, Münster, Hamburg, Saarbrükken, Kaiserslautern, Berlin, Kassel, Würzburg, Freiburg and Marburg. Technical universities too are beginning to focus more sharply on this field of studies. In addition to the aforementioned institutes, the strongly diversified R&D system in Germany also includes other establishments involved with nanotechnology, such as the NMI in Reutlingen, AMICA Aachen, IMS-Chips in Stuttgart, FBH Berlin, Bessy II Berlin, PTB Braunschweig, CAESAR in Bonn and IPHT in Jena.

Funding for Nanotechnology

Not counting the industry's own contribution, Germany's public expenditures for nanotechnology funding in 2004 total about ¢ 290 million. This does not include the Federal States' expenditures on the universities' basic budgets, nor the industry's own funding of nanotechnology research apart from public funding. To remain successful in the face of increasing globalisation, Germany must be conscious of its strengths in business and science and make better use of these assets. Nanotechnology's new potential for innovation is increasingly accessible thanks to interdisciplinary perspectives that span a range of industries. And the result is a qualitative leap for the technology's application and continuing commercial use, which must now be supported by decisive moves in research policy. This is where the development of products and systems based on nanotechnological advances and the integration of nanostructures in micro- and macroscopic environments presents an opportunity that must not be missed.

In many areas of nanotechnology, Germany is still out in front of many other countries in terms of knowledge.

This advantage, when paired with the production and sales structures needed for implementation and the internationally renowned German talent for system integration, must consequently lead to success in the marketplace. And this is exactly the field of application for the planned innovation initiative—"Nanotechnologie erobert Märkte" (nanotechnology conquers markets)—and for the new BMBF strategy in support of nanotechnology, which is being developed for the initiative. Until now, aspects of nanotechnology have been advanced within the confines of their respective technical subject areas.

However, the primary aim of incorporating them into an overall national strategy is to build on Germany's well-developed and internationally competitive research in science and technology to tap the potential of Germany's important industrial sectors for the application of nanotechnology through joint research projects that strategically target the value-added chain. This development is to be supported by government education policy to remedy a threatening shortage of skilled professionals.

To realise that goal, forward-looking political policymaking must become oriented to a uniform concept of innovation, one that takes into consideration all facets of new technological advances that can contribute to a new culture of innovation in Germany. And that includes education and research policy as well as a climate that encourages and supports innovation in science, business and society. Germany's economic future largely depends on how determined the country is to seize the opportunities presented by new technologies such as nanotechnology and on how successful it is in transferring those opportunities into economic gains.

In order to ensure that the society can enjoy growth, employment and prosperity for the mid and longterm, therefore, advances in research must consistently be applied to those areas where there is the most dynamic change and where market-leading innovations are possible. For many of Germany's important industrial sectors including the automotive business, IT, chemistry, pharmaceuticals and optics—the future competitiveness of their products depends on opening up the nanocosmos.

The share of the 2002 gross domestic product generated by the chemical industry, mechanical engineering, electronics, IT and communications technology and automotive manufacturing was approximately 36 per cent. With export quotas that can exceed 50 per cent, it is exactly these economic sectors and their innovative products that are Germany's top earners on the global market.

In contrast, the export quota registered by the industrial sectors that are not research intensive was only about 25 per cent. Key technologies such as nanotechnology thus directly influence the potential for growth and employment in the German economy's largest sectors, which must be used as the safeguards of the nation's prosperity. Technology and innovation are the decisive factors that are having a growing impact on the ability to compete with low-wage countries.

How greatly the prosperity of our society relies on technological advances—the development and use of key technologies—is illustrated by an example from the United States: More than two thirds of American private assets were created after 1970. The largest companies in the world, based on their market value, are all less than 30 years old, and the majority of them are technology companies that produce and market innovative products. It is now plain to see that the dynamic growth of nanotechnology has begun in earnest, and the German government's research and innovation policies have made a vitally important contribution to this development.

Those policies include an energetic commitment to research and development that has combined with dramatic increases in research expenditures by businesses to once again drive the research-and-technology share of the gross domestic product to 2.5 per cent. Now it all depends on setting a proper course for the future support of nanotechnology and to organise its breakthrough as an economically viable technology.

Market and Employment

The creation of jobs with secure futures is the central responsibility of domestic politics, and the reserves of all political forces must be mobilised if this responsibility is to be met. Besides introducing job-market reforms, new forces of growth must be enlisted in the effort. Research policy, in particular, can point the way to more growth and help create jobs that have mid-range and long-term futures. The key to this push is to vigorously channel research funding towards innovation. Innovations form the foundation of Germany's competitiveness and, as a result, the basis for growth and employment in our country.

In the age of globalisation, innovations are the lifeblood of our economy. They keep the economy moving, offset lost jobs and create prosperity. In the age of globalisation and growing communication networks, knowledge is available around the world. But the only people who will be creating jobs and enjoying growth are the ones who are the first to implement innovations.

Granted, it is difficult to quantify a relationship between innovations in new technologies and the employment system. As a result, it is not possible to make a simple forecast about the effects on the job market in a national economy. But no one doubts that global shifts in the job market are occurring between national economies that have received an innovation boost and those that have not.

Rough estimates show that the global market volume affected by nanotechnological knowledge is around ¢100 billion. That corresponds to approximately 500,000 jobs. Among the sectors with the highest job totals, there are at least three whose futures will be decided in part by expertise in nanotechnological processes. These three sectors -information and communication technology, motor-vehicle technology and chemistry—employ more than 2 million people in Germany. The international competitiveness of these sectors is based largely on knowledge and research intensive industrial areas that are forced to use technical innovations in their effort to adapt faster to demand than their competitors.

Increasingly, nanotechnological knowledge is flowing into other R&D-intensive sectors, including optics, biotechnology, medical technology and measurement engineering. This knowledge will play a competitively decisive role in the future positioning of products. In such areas, the newly redirected nanotechnology funding efforts by the BMBF will provide fresh momentum in the drive to launch market-dominating innovations and to initiate sustainable mid-range and longterm opportunities for growth, employment and prosperity in Germany. Given the positions being taken on technological capacities, the necessary contributions to the creation of jobs with secure futures and the current changes in innovation processes, an intensified research-policy focus on new technologies is today paramount.

Technological mega trends such as miniaturisation, individualisation and networking are playing a major role in the direction being taken in research policy. The technological mega trend of miniaturisation can illustrate the potential of nanotechnology as a future key technology to delve into dimensions never reached before and to open the way for the creation of products that are equipped with considerably improved or totally new functions. At the same time, social changes such as an increasingly aging population or the desire for mobility and security must be considered. The BMBF is taking on this challenge and is shaping its innovation policy in the area of nanotechnology funding accordingly.

The primary goals of this effort are to enhance the economy's profile in global competition, to consolidate and extend economic strengths, and to seize upon new developments in technology, the economy and society. A higher-profile effort to promote research in the area of nanotechnology should focus primarily on areas where special economic leverage can be exploited. This would include creating jobs with secure futures, preserving or expanding technological leadership, integrating ranges of services, and supporting German companies as system leaders in the global market.

For this reason, it is imperative that the lead markets in Germany should become the centre of attention, particularly in the areas of automotive manufacturing, machine construction, optics and chemistry. One key aspect of these lead-market sectors is their particularly close relationship with the German scientific community, from which they obtain their technological strengths. This relationship must be supported with respect to nanotechnology in future.

The task of researchers and politicians is to seize the opportunities of this new technological field by strengthening Germany's lead markets and by opening new markets in the drive to tap nanotechnology's potential to solve such social problems as unemployment and sustainability.

Furthermore, it is also important for the BMBF's funding policy that the questions concerning the possible applications of nanotechnology and their consequences which are being intensively, and sometimes controversially, debated in the public arena are faced and answered. Now that some areas of nanotechnology have matured, we can and must devote more effort to seeking a balance between public research and the strategic interests of German industry.

In the process, we must concentrate on the strengths of the German economy in order to achieve the maximum possible leveraging effect in economic terms. Research efforts are being focused on key areas of innovation, i.e. on strategic technological developments pursued jointly by industry and the scientific community with a pooling of research capacities and funds across multiple technologies. These partners are founding strategy-orientated research projects that can foster "leading-edge innovations" which highlight the economic possibilities and the social utility of nanotechnology.

Unfolding their considerable economic potential, these innovations are intended to exert the maximum leveraging effect on growth and employment along the entire value-added chains. In particular, they should "strengthen strengths." This involves the strengthening of nanotechnology itself as a driver of growth in many - particularly the lead market—sectors, its increasing connectedness with other technologies and its integration into applications, the consolidation and expansion of existing markets and the development of new growth fields.

Addressing the basis for innovative strength, this innovation strategy aims to secure competence in cross-disciplinary technologies such as nanotechnology as enablers for application-orientated product and system innovations, and to identify innovative fields of technology with immediate applications and cultivate them systematically where lasting differentiation and competitive advantages can be attained in global innovation processes characterised by a division of labour. Industry is called on to lead projects of this kind and participate in them to a significant degree. Leading-edge innovations are supported as collaborative projects between partners from the scientific community and the commercial world.

The BMBF will publish funding announcements concerning each of the leading-edge innovations. Leading-edge innovations dealing with socially relevant applications aspire to become trendsetters for economically successful BMBF funding measures and to open up new avenues of more rapid realisation. In strong sectors, all of the participants needed to generate added value are usually present in the country, so that a closed chain of implementation can be described and supported.

A complete description of leading-edge innovations also requires working out roadmaps that link the technical possibilities with the market conditions and the necessary strategy so that, starting from the social utility and market potential of ar innovation, one can derive the requirements for research and development, establish the necessary work packets and schedules on that basis, and describe the necessary contributions of the partners involved, including the required activities in various technical programmes of the BMBF.

Concrete projects that then result must be understood as elements on the road leading to a successful outcome over the long term. A scenario based on a leading-edge innovation must include descriptions of elements such as networking, early agreement concerning the goal to be achieved and the broad, added utility of the results. Those involved in nanotechnology are becoming increasingly interested in orientating basic nanotechnological developments to a broad range of potential applications in order to participate in the worldwide markets.

Numerous discussions with individuals from the scientific community and industry have resulted in the following examples of fields of innovation that aim to extend the lead acquired in the area of nanoelectronics, integrate nanotechnology into core sectors of the German economy, such as the automotive industry (NanoMobil), develop new areas of application (NanoLux), and enable interdisciplinary approaches (NanoLife).

Nanoelectronics

Electronics represents the foundation of nearly every innovative technology of our time. There is now hardly any piece of technical equipment that can do without electronic components. In Germany alone, the market for electronic components has a volume of approximately ¢ 20 billion and employs over 70,000 people. Even more important, however, is the leveraging effect that electronics has on innovation because of its fundamental character. Worldwide, the electronics industry is already the leader among the manufacturing industries, having already surpassed even the automotive industry.

Of particular importance, however, is the fact that no other manufacturing sector generates as much added value as electronics. This accounts for the special character of its key position in economic terms, too. For those who want to position themselves at the forefront of the world market with innovative technological products, electronics is an indispensable element of the value-added chain, an element that will continue to grow considerably in importance.

Leading-edge innovation

One example of a leading-edge innovation the BMBF has already taken up is the measure to support next-generation nanoelectronics manufacturing methods. This includes the project support for EUVL lithography—funded since early 2001 with over ¢ 50 million from BMBF over a period of 5 years—which lays the groundwork for future microchip production on the basis of powerful hardware technology.

The industry expenditures of over ¢ 120 million represent more than double the amount of this government funding. Another central area of support is the collaborative project "Imaging Techniques," in the framework of which the first leading-edge mask centre in Europe is bringing together the strategic advance research for chip production. The BMBF will provide approximately €80 million for research into mask technologies and their alternatives for manufacturing nanoelectronic chips with feature sizes ranging from 65 nm down to the ultimate limit.

The fabrication of these nanoelectronic chips uses a variety of different nanotechnologies. Reflecting layers with a precision on the nanometre scale allow the controlled exposure of masks. Pre-calculated nanostructures can offset diffraction errors, which makes it possible to write features that are less than half the size of the wavelength of light. Optical image distortions can be compensated for through nanoscale structures and layers.

Mask substrates must be atomically flat over large areas. In an industrial maximumthroughput process, nanoparticles must be found and removed to enable extremely low fault levels. The nano-mechanical influences of gravitation must be offset. Heavy, solid machine parts must be positioned with nanometre precision and minimum delay. All of this serves to make available nanotechnology with an extremely high level of productivity.

Nanoelectronics is therefore the first technology to penetrate the nanocosmos on a broad front—not just in selected niches but especially in mass production for standard applications ranging from the desktop PC to the automobile—and the first to fulfil the expectations people have of nanotechnology in a tangible, affordable way, in the form of a marked increase in performance at steadily falling prices. The electronics industry is the only technology sector that has formulated the development of nanoscale technologies in the form of a roadmap which is systematically followed as a guideline by all the leading chip and hardware manufacturers worldwide. The automotive industry is thus one of the most important drivers of the German economy. Its share of global production comes to 23 percent when mergers are taken into account.

There are approximately 730 million cars on the road worldwide. In China and India alone, the demand for mobility is estimated at about 600 million cars. To a great extent, the success of the German car manufacturers is essentially due to the domestic suppliers. The latter consider their greatest challenge to be the stabilisation of their technological expertise, which is crucial for the German market position. The German manufacturers have opted for a strategy of using top-class technology to gain a competitive edge, which creates the need for large R&D expenditures.

Attractive products and new market opportunities can be attained only through the creative joint efforts of all those involved in the process of creating value. In order to strengthen these strengths of the domestic admittedly still no proof that structures of this kind can be manufactured economically and hence made accessible to the public at large. However, the history of electronics shows that man's inventive genius has so far always been able to overcome any obstacle on the road to progress in electronics industry, it would be appropriate for the participants to combine their efforts in the framework of a leading-edge innovation.

NanoMobil

According to a study conducted by the Office for Technology Assessment of the German Bundestag, nanotechnological expertise in the automotive design of the future is one of the core capabilities that are absolutely essential for maintaining international competitiveness. The automobile is used for both work and leisure; it ensures individuality and guarantees mobility.

There are approximately threequarters of a billion vehicles on the road worldwide, and the number is growing. Hence, the problems of vehicle consumption, pollution emissions, recycling and the volume of traffic also represent opportunities for industry, particularly for the German automotive industry, which is so strong internationally.

This is a lead market, and its strength is, on the one hand, its role as a driver of technology, because a leading edge innovation of this sort can open up prospects for applications in other fields. In addition, its strength comes from its role as an important customer of the supplier industry, which is dominated by medium-sized firms. A leading-edge automotive innovation that uses nanotechnological expertise for new functionalities will aim to optimise sustainability, safety and comfort.

Making progress toward the leading-edge innovations' goals of creating jobs and encouraging sustainable developments requires new solutions derived through nanotechnology, materials research and adaptronics, in particular. With a view to possible paradigm shifts in future automotive design, the leading-edge innovation NanoMobil must therefore make an extra effort to involve R&D institutes and innovative suppliers that develop basic discoveries in nanotechnology, that evaluate them with respect to their use in the car of the future and additional broad-based industrial exploitation, and that realise them in new products together with suitable industry partners.

Life sciences and nanotechnology

In addition to strengthening the already strong sectors of industry, other aims of leading-edge innovations are to tap into long-term market potential at the appropriate time, create new products and services, and be the first to participate in the global sales opportunities. Future market opportunities of this kind can already be anticipated in the pharmaceutical and medical fields, hence the advisability of cross-disciplinary activity in the life sciences and nanotechnology.

The biotechnology sector has been experiencing dynamic growth worldwide since the mid-90s. The USA is still the leader in the field, ahead of Europe. This edge is not a function of the number of companies involved. Instead, it is a reflection of the maturity of the sector in the USA. The German biotechnology scene can now likewise point to many innovative biotechnology companies that have successfully positioned themselves on the international market. Of the 4,300 biotechnology companies around the world, 360 are located in Germany.

These are predominantly small and medium-sized companies with fewer than 50 employees. The biotechnology sector is characterised by large investments in R&D. In 2002, the roughly 600 listed companies, which operate mainly in the field of red biotechnology, invested approximately US $22 billion in R&D and had total sales of US $45 billion.

The biotechnology sector has come to be viewed as an engine of innovation for the development of new therapeutic and diagnostic articles with the help of which the large pharmaceutical companies, too, will be able to fill their product pipelines and expand their product portfolios. In order to encourage a similarly successful take-off for nanotechnology, the founding of innovative start-ups and the growth of these companies must be promoted at an early stage. The BMBF will support this process with supporting measures. The integration of small and medium-sized nanotechnology firms into value-added chains not only makes an essential contribution to technology transfer, however. It also makes it possible to tap into new fields and markets with a high potential to generate added value.

Nano for life

The aim of the leading-edge innovation "Nano for Life" is to make a decisive contribution to the health of society through the increased use of technologies and insights from the fields of nanomaterials research and nanobiotechnology. This requires, on the one hand, a strengthening of the growing nanotechnology sector in Germany, the innovative power of which depends to a great extent on research-intensive SMEs.

These small and medium-sized companies can aptly fill specific niches in the early phase of value-added chains that focus on the development of medical products and pharmaceuticals. Furthermore, the sectors in Germany that are traditionally strong and are relevant to the health market, such as pharmaceuticals, chemicals, biotechnology, and material and medical engineering, can profit from the introduction of nanotechnological processes and applications by generating completely new products with considerable socio-economic utility.

On account of the demographic make-up of the population and the availability of innovative new medicines, health expenses in Germany are continually rising. In 2001, these expenditures came to ¢ 225.9 billion, or 10.9 percent of GDP. For the most part, experts agree that nanotechnology will open up new prospects in the development of therapeutic and diagnostic products. But its considerable potential with respect to resource conservation could also help to keep health costs down over the medium term. The availability of highly sensitive diagnostic products improves the chances of prevention and early treatment of serious illnesses. The use of DNA or protein arrays, for example, makes it possible to identify those patients who are very likely to respond to a certain medication. This procedure not only cuts costs, it can also lead to a reduction of drug side-effects.

Nano-medical applications include nanoparticles that act as drug-delivery systems, or are used for hyperthermal treatment of tumours or as vectors in gene therapy. Nanoparticles likewise play an important role in some imaging techniques used in medicine. In addition to nanoparticles, there are the interesting present-day nano-applications of biosensors and nanostructured implant coatings. Here, markets with considerable sales potential can be tapped in the short term, since the demand for artificial replacement tissue, far exceeds the supply of donor materials.

8

Public Concerns about Nanotechnology

Emerging technologies such as nanotechnology offer new opportunities and new dangers, posing questions for researchers, ordinary citizens, government officials, and others about what roles they may play in promoting safe and socially beneficial development and application of new technological capabilities. Nanotechnology, the precision control of materials at the scale of individual atoms, represents a set of technologies that could have profound impacts on all aspects of life. Atomic force microscopes, such as those used to place individual xenon atoms spelling out the IBM logo, are part of a growing capability to observe and manipulate individual atoms and molecules at the nanoscale of billionths of a meter. Nanotechnology products currently on the market have just begun to enter a few specialized applications, but large investments, already in the billions of dollars per year, are being made in nanotechnology research and development. Advances from microelectronics to nanoelectronics hold the prospect of further shrinking and accelerating computers and other information technologies, as well as offering much more efficient ways to trap and hold solar energy.

Nanotechnology is expected to enhance and complement biotechnology, developing sensors, precision drug delivery systems and nanomedical devices to make repairs within cells, improving health and extending life.Nanotechnology should offer greatly enhanced capacity to prevent and clean up environmental pollution, for example filters to purify water and extract pollutants. Nanotechnology is expected to yield better materials and products, fashioned more precisely, less wastefully, and at much lower cost. There are technical debates about the feasibility of specific forms of nanotechnology, particularly the "molecular nanotechnology" of self-replicating assemblers proposed by Eric Drexler in the 1980s. Drexler, and the Foresight Institute he helped establish, have been concerned with the possible benefits of nanotechnology, and even more concerned about the dangers it might pose, and the need to prepare society for powerful new technologies.

However even leading scientists who disagree with bold visions of nanobots, global abundance of ultra-cheap manufactures, and rejuvenation, still argue that nanotechnology can be expected to have dramatic consequences for energy (e.g. cheap solar power), health, manufacturing, information technology, space travel, and augmentation of the physical and mental capacities of human beings. Most discussion about the potential dangers of nanotechnology has focused on the dangers of accidents with self-replicating nanomachines, "gray goo"or use of nanotechnology in warfare, but there has been much less discussion of the broader social and economic impacts.

Apocalyptic threats, of accidents and abuse have attracted much attention However, the most important risks and opportunities are likely to come from more prosaic applications that drastically reduce costs and increase capabilities, a disruptive general purpose technology with profound consequences for economies and societies.

One possibility is that the public is ignored and excluded from discussions about research and development.

Technical complexity makes it convenient to confine discussion to suitably qualified experts. However, experience from other fields indicates that this could lead to major problems downstream. The story of nuclear power in the United States is the most conspicuous example of how overconfidence in the efficacy of safety measures, secrecy, and exclusion of the public ultimately bred distrust and polarization, creating "technological gridlock."

Simplistic discussion focusing only on the upside of nanotechnologies would be similarly likely to breed suspicion, comparable to earlier advocacy of nuclear power "too cheap to meter" from "the friendly atom." Ignoring the public may be an easy option in the short term, but creates the risk of criticism and major adverse impacts on reputation and business prospects later on. In Britain and other parts of Europe, controversies about "mad cow disease" are another example of the loss of public trust in regulatory agencies dealing with complex scientific problems. Blanket claims that nanotechnology products are "substantially equivalent" to existing products, and so do not deserve any special labeling or regulatory attention risk running into the kind of controversies which have surrounded genetically modified foods.

In the U.S., the theory of substantial equivalence permitted introduction of GM foods with relatively little public discussion for a period of time, but was later followed by alarms about "Frankenfoods" and vigorous opposition to GM foods. Secrecy, silence, and overconfidence in the potential for containment and accident prevention may breed later problems. Experience suggests that all organizations involved in nanotechnology research and development, including government agencies, research institutes, and businesses, would be wise to undertake a pro-active approach to engaging the public.

Sharing Information

The first step on a ladder or spectrum of participation is sharing information. Nanotechnology is now receiving increasing coverage in newspapers and magazines, expanding beyond more specialized technical communities and science fiction stories. Most research has been conducted through science and its largely open procedures, and published in scholarly journals, through now patents and trade secrets are beginning to play a larger role.

Nevertheless discussion is complicated by technical complexity. Aside from gray goo of swarming nanobots, there are still few easily digested sound bites for media. For most people, rejuvenation through nanomedicine, use of structural diamond as a commonplace building material, or household nanofabricators making products to order as easily as photocopies are currently printed, are still visions far too speculative to be taken seriously. Many existing institutions already promote availability of information about some nanotechnology research and development.

Government agencies, such as those funding nanotechnology research, are generally required to share information with the public about their activities. Interested citizens, journalists and others can also request more detailed disclosure of official records. Private companies have fewer obligations to share information, but are still required to file certain kinds of information with regulatory bodies such as the Securities and Exchange Commission and patent offices. To the extent that products, such as titanium nanoparticles in sunscreens and "Nanocare" antistain treatments on clothing, explicitly advertise their presence, such labeling may reduce worries about invisible and involuntary exposure. Public and private organizations can also employ the usual range of public relations techniques to distribute information through press releases, briefings, interviews, tours, e-mail lists, webpages with Frequently Asked Questions (FAQs), advertising and other methods.

They may also produce technical reports, journal articles and other documents aimed at more specialized audiences. Growing public concerns suggest that companies, research institutions and other organizations involved in nanotechnology research and development may wish to do more to make information available, and to engage in additional measures to improve understanding and discussion about how nanotechnology can be wisely developed.

Consultation

Consultation involves a shift from one-way communication to two-way interaction, from education to dialogue. Typical consultation methods include public hearings, focus groups, suggestion boxes, open houses, and question and answer sessions. Broader discussion in the media, policy studies and other forums can also be considered a form of consultation that offers input into societal decisions about nanotechnology.

Legislative hearings on nanotechnology have provided some opportunities for comment from invited experts. However until recently public discussion has largely been confined to relatively small groups of aficionados,and a few early warnings from groups concerned about negative impacts of technology. Despite some relevant precedents in the human genome initiative, relatively little attention has been paid so far to social and ethical implications of nanotechnology.

Resources are now becoming available, for example funding from the U.S. National Nanotechnology Initiative, for research on social and ethical issues. These could be employed to improve public participation in nanotechnology research and development. Since the 1980s, the Foresight Institute, founded by Eric Drexler, has sought to promote discussion of how to deal with both the promise and perils of molecular nanotechnology. Foresight has produced newsletters, papers, books, and a website to share information. It has organized conferences that have concentrated on technical issues but also included some papers dealing with issues of safety, international security and other broader implications Foresight has tried to engage in dialogue, both with those questioning the feasibility of nanotechnology and others alarmed by the dangers it might bring.

Another recent attempt to promote better informed public discussion comes from the Center for Responsible Nanotechnology, which is preparing a series of papers on various issues. For example, their analysis of how interpretations of precautionary principles might be applied in nanotechnology argues strongly for weighing the benefits that might be foregone if new technological capabilities are not developed.Risk communication issues pose a major challenge for public discussion about nanotechnology. Several "outrage" factors likely to elicit concerns about riskare present, including:

— Novelty
— Invisibility
— Involuntary exposure
— Dread of catastrophe

Public concerns about risks of GM foods were exacerbated by the possible invisible and involuntary nature of exposure, issues which will also be present with nanotechnologies. Assertions about the feasibility of containing the spread of genetically modified organisms were undermined by controversies about dissemination of pollen from genetically engineered corn into nearby fields, and by the appearance of Starlinkcorn, intended only for animal consumption, in tortillas.

Arguably, GM foods offered few clear benefits to consumers, aside from possibly lower prices, even though they did offer benefits for producers in reduced costs for pesticides, which might also bring positive environmental benefits. The scale of nanotechnology makes it "invisible," and increases the likelihood that exposure may be involuntary.

Scenarios of runaway self-replicating nanomachines generate dread of catastrophic outcomes. Some of these risk concerns can be addressed by ensuring control and containment, and voluntary, informed use of the technology, issues which are already being addressed in fields such as regulation of toxic substances, protocols for genetic therapy and safety measures in medicine using radioisotopes.

Nevertheless these characteristics are likely to generate continuing concern about risks. Alarms about apocalyptic scenarios of accidents and abuse may also divert attention from the broader implications of emerging nanotechnologies. Drexler and others concerned about nanotechnology have sought to point out potential benefits such as improved medical care, environmental protection, and poverty alleviation, making the positive benefits for ordinary people much clearer.

Consultation tends to mostly involve those already interested in public policy issues. While discussion in mass media often highlights dangers, and concentrates on simple sound bites, some of the issues involved in nanotechnology research and development deserve deeper and more balanced consideration.

Public Involvement

Public involvement programs go beyond information and consultation to systematically engage stakeholders in identifying their concerns and ensuring they are addressed in the final decision. Workshops and other forums can offer opportunities for more intensive dialogue, the results of which can then be used by those making the decision. A good process should provide the public with feedback on how their inputs affected decisions. In collaboration, through citizen panels, task forces, advisory committees and other methods, stakeholders have a "seat at the table" enabling them to participate actively in developing alternatives and identifying preferred solutions.

For public participation at the level of involvement or collaboration, the final decision still lies in the hands of a regulatory agency, private company, or other ultimate decisionmaker. Typically, such a decisionmaker (or series of decisionmakers in a government or business hierarchy), acts on the recommendations formulated by a collaborative group in which stakeholders have a major role.

However it is worth noting that stakeholders have not only voice in the process, but power in the extent to which they retain the option of resorting to other channels, such as criticism in the media, political lobbying, and litigation in court, if they are not satisfied with the direction or the results of the process. So far for nanotechnology, little use has been made of more systematic methods for promoting discussion, such as stakeholder panels or citizen juries.

In dealing with technically complex issues, these methods have the key advantage of allowing a representative group of people time to learn about and consider new, technically complex issues, moving beyond first impressions to deeper examination. Citizen juries and similar methods have been used to examine questions in biotechnology as well as other social policy issues. Stakeholder panels have been employed for activities such as deciding the location of new facilities and developing emergency plans for chemical plants.

In the case of nanotechnology, such public involvement methods could help to overcome some of the barriers posed by technical complexity. They might also help direct attention to the broader implications of nanotechnology for everyday lives, redressing the bias toward attention to catastrophic scenarios of accident and abuse that are more attractive for media coverage. A prominent example of collaboration is negotiated rulemaking, where regulatory agencies involve representatives from potentially affected businesses, concerned civic organisations, specialist experts, and others to take part in drafting regulations.

Such methods can draw on knowledge of the participants and formulate regulations that are likely to be more feasible to implement, and less likely to result in objections and legal challenges, than regulations formulated by an agency unilaterally. Negotiated rulemaking could play an important role in the development of regulations concerning nanotechnology, particularly given the technical complexity of many of the issues. Regulatory attention to nanotechnology is in an early stage. However, much research is already guided by existing safety procedures, and products would come under existing regulatory regimes.

A key question thus concerns the extent to which new or special measures are needed to deal with nanotechnology. Recently raised concerns about research and applications involving carbon nanotubes and other nanoparticulates, and advocacy of a moratorium until consensus is developed on suitable safety precautions in nanotech R&D, is likely to be only one of many debates about the extent to which nanotechnology deserves special regulatory attention.

The nature of public involvement will depend on the kind of regulation that may be pursued. For the case of self-replicating assemblers, molecular nanotechnology, there are special challenges. It seems likely that prohibition is unworkable, and military monopolization of nanotechnology research and development undesirable.If so, then civilian research and development might be guided by an approach similar to that used for recombinant DNA technology. This could rely primarily on selfregulation and responsible professional practice, backed by a modest level of government regulation. If such an approach were able to forestall accidents and develop the capabilities needed to defend against abuse, then the other major issues would concern the impacts of molecular nanotechnology as a new general-purpose technology with major impacts. A broader issue concerns the allocation of effort between attempting to define and anticipate risks in advance and developing capacity to resiliently respond to risks over time, particularly in cases where there is a large degree of uncertainty about how new technologies will actually be applied.

Rather than only devoting resources to planning and anticipation to prevent problems, it may be more productive to balance anticipation with resilient strategies for learning from and responding to problems as they actually appear.Typically in the discussion that occurs as part of consultation or public involvement activities, participants bring their own broader agendas, for example proponents seeking to profit from a technology, politicians pursuing publicity and support, and nongovernment groups with their own particular missions. The recent calls for a temporary moratorium on nanotech development come from a group that has been active in debates about genetically modified organisms (GMO), particularly new biotech crops.

They are extremely concerned about international inequities in access to new technologies and the dangers of corporate monopolies exploiting and abusing intellectual property rights. Regulations focus on decisions made through government procedures. A strong government role may be advocated as a way of promoting democratic participation in decisions about development of technologies with major implications for societal welfare.

A key issue concerns the extent to which decisions are made through a centralized regulatory process, versus being left up to more decentralized actions of individuals, corporations and others acting on their own, though still possibly constrained by broader regulatory restraints and guidelines.

Empowerment

The next level of participation is often discussed in terms of empowerment. Empowerment offers participants greater ability to determine their own futures, in contrast to just being informed, listened to, engaged in participatory processes or carrying out joint decision-making. Participation need not be restricted to citizens taking part in government actions to legislate, plan or regulate, or working in partnership with government, but includes the space for individuals, groups and communities to act autonomously and independently. Empowerment is a broad category that can be further subdivided into several different levels: Partnership (co-management); delegated authority; regulated autonomy; responsible independence

Under public involvement and collaboration, the final authority remains with one party, such as a government agency or private company, even if that decisionmaker informs, listens to, and involves others in the decision process. By contrast partnership means all parties take part in determining decisions, so that authority and responsibility are genuinely shared.

Decision rules might require unanimity, where all parties have a veto; consensus, where everyone agrees to support, or at least not oppose, an outcome; or be based on a majority vote or other procedure. When an agency of government is involved in such joint decisionmaking the power relationship is inherently asymmetric, but is still significantly different from a situation where government retains final decisionmaking authority.

In some cases the pursuit of consensus in negotiated rulemaking may approach a partnership or collaborative approach. Other participants willing to engage in legal challenges to rules with which they are unhappy have a strong weapon to bring into play, even if the immediate authority to issue regulations lies with government.

Where government agencies enter into cooperative arrangements with communities, private sector groups, or other entities, with both sides having a strong say in framing the terms of coöperating and deciding whether or not to act together, then the arrangement can be seen as one of partnership. One area where public-private partnership arrangements might play a significant role in nanotechnology lies in establishing relevant standards, e.g. for software and communication interfaces in nanotechnology, as an alternative to proprietary standards controlled by a single company, or standards unilaterally established by government.

Where a concerned community takes substantial initiative in proposing rules, while inviting government to provide authoritative sanction for the final rules, then the relationship goes beyond collaboration to be more one of partnership. The process by which regulations for recombinant DNA were developed is a good example in which the scientific community took the initiative to develop rules for promoting safety in research and development of an emerging technology.

Scientists, engineers, and others concerned about preventing accidents and reducing the risks of abuse of technology need not wait for government regulation, but can also take initiative themselves. A further level of empowerment allows stakeholders the freedom to make decisions themselves, within regulatory guidelines and reporting requirements.

Beyond specific regulatory requirements, a community, business or other entity which is legally liable to harm created by its products will need suitable insurance, and so have incentives promoting responsible development of nanotechnology.

The potential legal consequences for product liability would help to regulate behavior, promoting safe use and providing incentives to prevent accidents and abuse. To the extent that companies seek to obtain insurance to protect themselves against liabilities this could stimulate sophisticated assessment of risks and other problems, with a minimal government role. In addition to specific worries about liability, individuals, corporations, and other institutions, such as universities, may also be motivated by concern to protect their reputation.

Consequently, they have incentives as private actors to respond to social concerns. Liability and reputation offer a way of promoting responsibility in how individuals and corporations use their independence. They also offer a source of leverage for potential affected individuals and groups, and other activists, which may bypass or complement more conventional regulatory routes for dealing with risks and social impacts of new technologies.

These dynamics have clearly been at play in debates about GMOs, evaporating the value of stocks in companies involved in agricultural biotechnology and discouraging further investment in the field. Reputational incentives may also play a significant role in nanotechnology. Participation in nanotechnology also involves issues of who has access to knowledge and who has rights to develop the technology. Intellectual property law concerning patents, copyrights, trade secrets and related issues has begun to play a major role in affecting access to technology. This invokes issues about the extent to which patents offer suitable incentives for innovation and the extent to which they may needlessly promote monopoly, exploitation, excessive transactions costs, inequity, and other problems.

The application of "open source" approaches to software development, as exemplified by the Linux operating system and the development of hypertext markup language for the worldwide web illustrates one way of expanding participation.

Application of such approaches to software and hardware for nanotechnology could make a major difference in who gains access to nanotechnology.Such approaches may promote applications that empower individuals to use and develop the technology. Conversely, highly restrictive licensing could promote monopolies. One of the challenges in promoting public discussion about empowerment and self-governance in nanotechnology R&D may lie in how to avoid polarization and simplified rhetorical stereotypes of total laissez-faire and extreme over-regulation, and instead engage in a more nuanced analysis of issues and alternatives for technology governance. While empowerment is unlikely to be the only strategy involved in the governance of nanotechnology, partnerships, delegated authority, regulated autonomy, and responsible independence may still supply an important complement to consultation, involvement, and collaboration in government regulation, as part of developing a suitable system of technological governance.

In UK, technologies, such as genetically modified (GM) crops and food, public attitudes play a crucial role in the realisation of the potential of technological advances. A single quantitative item appeared on the 2002 Eurobarometer survey, where over 50% of the sample answered 'don't know' when asked whether they thought that nanotechnologies would improve or make worse their way of life over the next 20 years.

Of the remainder who did have an opinion, a clear majority felt that it would indeed improve their lives. However, the extremely high level of 'don't know' responses indicates very low general levels of awareness of the issue of nanotechnologies across Europe. A webbased survey conducted in 2001 in the USA jointly sponsored by the National Geographic Society and the National Science Foundation found that 57% of respondents agreed with the statement that 'human beings will greatly benefit from nanotechnology, which works at the molecular level atom by atom to build new structures, materials and machines'. However, such a web-based sampling technique is inherently self-selecting in nature, drawing disproportionately from people who have ready internet access as well as those who are particularly interested in science and technology issues in the first place. Accordingly, it is impossible to extrapolate this result to attitudes among a sample of the general public.

A further difficulty with both surveys is that we cannot know whether the responses obtained reflect genuinely considered beliefs about nanotechnologies, or a response to the questions based upon beliefs about the future impacts of technology more generally (where attitudes are known to be highly favourable). That is, neither survey gives us detailed information on how people might interpret a new development when it is described to them in some detail, something that is arguably more important as an indicator of the way in which public attitudes to nanotechnologies might develop in the future.

In interpreting the findings from the survey and qualitative workshops, it should be borne in mind that both were exploratory exercises, conducted within the remit and financial resources available to the Working Group. They certainly should not be taken to represent a full exploration of current British attitudes to nanotechnologies. In addition, the findings should be interpreted in a British context only: no generalisations can be made from these data about public attitudes to nanotechnologies in other countries, particularly to cultural contexts outside Europe.

Several issues are, however, worthy of comment. The fact that awareness among the British population is currently very low implies that much will hinge upon how attitudes to nanotechnologies are shaped over the next few years. In addition, when attempting to define this new development, the survey respondents made reference to other technologies with relatively positive associations. This may well explain in part why the majority of those who could provide a definition also thought that nanotechnologies would improve the quality of life for people. By contrast, in the qualitative workshops, where respondents deliberated the issues in greater depth, responses were more mixed and at times touched also upon issues with more negative connotations. Four issues arising from the qualitative workshops can be placed in relation to what is already known about public perceptions of risk.

First, the need for informed and accessible commentary on, and consideration of, any long-term uncertainties associated with nanotechnologies. Uncertainty has potentially both positive and negative outcomes, as the workshop participants fully recognised. However, it is known to be a significant driver of public concerns about technological risks, particularly where doubts exist over future safety or environmental impacts. Uncertainty was identified as a key factor in some of the very first research into risk perception on nuclear energy, and subsequent studies of a wide range of risk issues.

Second, questions over governance of nanotechnologies. Like the concerns over long-term uncertainties, these issues are not specific to nanotechnologies, but arise in public discourse about many other technological issues. It can be helpful to separate governance issues into two strands. The first involves the role and behaviour of institutions, and their abilities to minimise unintended consequence and adequately regulate. Such questions are not, as Wynne points out, the product of a mis-informed or 'irrational' public. Rather, they are legitimate questions touching upon areas of very real potential risk, albeit ones that are inherent to the way organisations and regulation operate, and as a consequence sometimes difficult to represent in formal quantitative risk assessments. Nor should such questions be seen as the product of views that are anti-science or antitechnology.

Many people, as the current workshop findings also indicate, remain highly enthusiastic about the general impacts that science will have on their future lives. A second strand concerns the possible trajectories that nanotechnologies will follow as they develop: who can be trusted to ensure that these trajectories will be socially beneficial? Can the public play a role in determining which trajectories are realised? Such questions express genuine doubts that people have about the ethics, social uncertainties and future governance of the technologies. Such concerns are likely to be key ones that will arise in any dialogue process involving nanotechnologies.

Third, the enthusiasm that many workshop participants expressed for the possible ways that nanotechnologies would benefit their and others' lives. Perhaps not surprisingly, benefits are an important part (if not the only part) of the evidence that people weigh up, alongside perceived risks, when making a judgement about the acceptability or otherwise of a hazard that might impact upon them.

Fourth, BMRB also report that there was some mention in the workshops of ethical concerns over 'messing' with the building blocks of nature, in part in response to suggestions of scientists manipulating matter at the atomic level to create entirely new materials. An analogy was also drawn here by some workshop participants between nanotechnologies and GM. Both quantitative and qualitative risk perception research that this issue is one determinant of uneasiness over biotechnology and other issues such as radioactive waste and nuclear energy. It follows that, not only does the potential exist for nanotechnologies to be stigmatised by such general association, but also some specific applications are likely to raise significant ethical dilemmas.

Awareness of nanotechnologies is currently low among the British population. In addition, the workshops reported here represent the first in-depth qualitative research on attitudes to nanotechnologies in the published literature, as far as we are aware. Their findings, although limited, provide a valuable indication of some of the wider social and ethical questions that ordinary people might wish to raise about nanotechnologies both now and in the future.

The assessment and control of the impacts of nanotechnologies—as a highly technical and complex subject—should be an expert-led process, restricted primarily to the peer community of scientists and engineers within academia, industry and government. Several recent UK reports have recommended that scientists and policy makers engage in dialogue with interested parties about science and technology issues, risk and the environment. In this respect, the events surrounding the bovine spongiform encephalopathy (BSE) crisis in the 1990s marked a turning point in the way UK science policy and risk assessment practice is viewed.

The House of Lords Science and Technology Committee in particular, in its report on Science and Society, recommended 'That direct dialogue with the public should move from being an optional add-on to science-based policy-making and to the activities of research organisations and learned institutions, and should become a normal and integral part of the process.' Its rationale was that a crisis of trust had arisen in certain areas of UK science policy-making. To regain public trust, it recommended greater openness and transparency about science policy and scientific uncertainties. This assertion did not go unchallenged. O'Neill has argued that the evidence is not clear that the so-called crisis of trust is a response to greater untrustworthiness of officials in the UK: rather, many statements of mistrust might actually reflect a climate of suspicion, partly fed by media reporting of issues. Dialogue with a range of stakeholders about risks also holds an increasingly important place in the work of many of the new advisory and regulatory bodies set up in the UK in the wake of BSE, such as the Food Standards Agency, the Agriculture and Environmental Biotechnology Commission, and the Human Genomics Commission. Dialogue-based processes have also been extensively used for addressing environmental and risk decision-making across Europe and the USA.

The Royal Society is undertaking its own dialogue initiatives through its 5-year Science in Society programme. The aims of this programme are captured by the President of the Royal Society, in his 2001 Anniversary Address: 'Society needs to do a better job of asking what kind of tomorrow we create with the possibilities that science offers. Such decisions are governed by values, beliefs, feelings; science has no special voice in such democratic debates about values.

But science does serve a crucial function in painting the landscape of facts and uncertainties against which such societal debates take place'. The general case for wider societal dialogue about novel technologies, and with it greater openness about science policy, rests upon three broad sets of argument. Fiorino characterises these as normative, instrumental and substantive.

The normative argument proposes that dialogue is a good thing in and of itself and as such forms a part of the wider democratic processes through which controversial decisions are made. The normative argument suggests, in particular, that it is important to make decisions sensitive, as far as is possible, to the ethical and value concerns of directly affected groups or populations.

The instrumental argument suggests that dialogue, as one means of rendering decision-making more open and transparent, will increase the legitimacy of decisions and through this generate secondary effects such as greater trust in the policy-making process. Many of the arguments in the 2000 House of Lords Science and Society report focus upon the issue of the legitimacy of risk regulation and science. Finally, the substantive argument is that dialogue will help to generate better quality outcomes.

In the field of environmental risk, non-technical assessments and knowledge have been shown to provide useful commentary on the validity or otherwise of the assumptions made in expert assessments. For upstream issues, where high levels of uncertainty exist, there may be particular benefits to opening up the risk characterisation process to a wide range of differing perspectives.

The aim here is to avoid an overly narrow framing of the problem, through giving consideration to as full a range of impacts as possible, including potential 'shocks and surprises', many of which may not, initially at least, be open to formal quantitative analysis. A US National Research Council report on Understanding Risk develops a detailed set of proposals for risk characterisation. They define the resultant analytic-deliberative process as combining sound science and systematic uncertainty analysis with deliberation by an appropriate representation of affected parties, policy makers and specialists in risk analysis.

Although there is clearly a considerable momentum in the UK and elsewhere to engage in dialogue over science and technology issues, this should not be viewed uncritically. A first challenge concerns defining who might be involved. A useful distinction in this regard can be made between 'stakeholders' and 'the public' as follows: 'the term, stakeholder refers to organised, official and defined interested parties in any decision, such as NGOs, environmental groups, industry, regulators. The term public refers to individuals and communities who have an interest or stake in an issue but who may be less organised and less easily defined and identified'. This distinction is particularly important when designing engagement processes, for who needs to be involved will depend upon the objectives of dialogue, and in turn will have a direct bearing upon the expected outcomes, their efficacy and legitimacy. For many issues even the category 'the public' should not be viewed as a single undifferentiated entity, particularly in terms of attitudes towards risk.

Other difficulties with dialogue processes arise because, as Okrent points out, we do not yet know enough about the practicalities and impacts of using analytic-deliberative processes. One reason for this is that systematic evaluation of dialogue processes and their outcomes remains relatively uncommon, being difficult and expensive to do properly, and often not recognised as important by sponsors at the outset of the process of dialogue. In addition, a number of technical and institutional/cultural barriers, such as regulatory fragmentation, may thwart effective implementation of an otherwise well-intentioned and planned dialogue process.

'Upstream' Issue

Most developments in nanotechnologies, as viewed in 2004, are clearly 'upstream' in nature. There are at least three senses in which this is so: regarding current decisions, impacts and public acceptance, respectively. First, many of the significant decisions that will affect the future trajectory of the technology, concerning research funding and R&D infrastructure, have yet to be made.

A timely and very broad-based debate might therefore focus upon which trajectories are more or less desirable, and who should be the ultimate beneficiaries of public sector investment in R&D, before deeply entrenched or polarised positions appear. Mehta argues that in the

Canadian context the failure to consult the public early over biotechnology has led to several difficulties in the regulatory process, while Mayer also points out that the problematic issues that heralded the advent of biotechnology in the 1970s and 1980s did not go away, and that the participative technology assessment methods only now being developed for biotechnology might be usefully deployed in the upstream phase of nanotechnologies.

Second, many of the social and ethical impacts of nanotechnologies are yet to be envisioned, remain hypothetical, or will depend upon nanotechnologies' convergence with other technologies. Only over the medium (5-15 years) or far longer (more than 20 years) term will its precise outcomes and associated ethical implications become clear. Achieving meaningful dialogue today will therefore set several difficult challenges: to separate current hype from what is realistically achievable with the technology; to provide good-quality information on likely impacts; and to scope fully the potential sources of uncertainty. In turn, very specific applications might raise unanticipated social or ethical questions only well into the future or when the technology has reached a mature stage of development.

Finally, in terms of public acceptance, the research presented above illustrates that nanotechnologies have yet to gain any major place in public discourse in Britain, with awareness of the technology among the general population being extremely low. Although this has the potential to change rapidly, research on the factors that lead to risk issues becoming amplified or attenuated in public discourse shows that this rarely depends upon any single factor operating alone. Rather, this depends upon the combined impacts of a range of factors accumulating over time. These include: the balance of perceived benefits between individuals, private and the public sectors; analogies drawn with other (both stigmatised or accepted) technologies;

patterns of media coverage; position of campaigning groups; the existence of significant scientific dispute; and attribution of blame for prominent 'accidents' were these to occur. Viewing nanotechnologies in upstream terms suggests that lessons can and should be learned from the history of other similar technological innovations. Mayer argues that the development of all major technologies should be viewed as social processes, and that framed in this way there are clear parallels to be drawn between nanotechnologies today and the position that biotechnology faced in the 1980s.

Similarities include the levels of excitement and hype, a promise to control the future without critical consideration first of which futures are desirable and who might ultimately control them, and narrowly framed debates about risk issues not encompassing wider social and ethical issues. One can also draw parallels between nanotechnologies today and the nuclear energy industry in the 1950s. Wynne argues that a particular difficulty with the early history of that industry, unanticipated at the time, was that the over-optimistic early claims made for the technology laid the foundations for the deep public scepticism and opposition that was to emerge much later in the 1970s.

Ultimately, it is difficult at this stage to judge whether, and which, applications of nanotechnologies will necessarily prove more or less difficult than nuclear power, GM or any other controversial technology. One can make the argument that with many of these more mature technologies public dialogue has typically arrived too little too late, only being seen as an optional 'add-on' when the decision-making surrounding an issue has become pressing, difficult or uncomfortable for regulators or governments. Under such circumstances the existence of highly polarised positions can make it very difficult, if not impossible, to take any real dialogue forward.

In broad agreement with, a number of submissions and papers that have argued for wider public dialogue and debate about the social and ethical impacts of nanotechnologies. However, the evidence presented to us also suggests that specifying the precise forms of such dialogue will be no simple matter.

The objectives of dialogue, alongside who needs to take part, are likely to vary over time as the issues with nanotechnologies evolve. Equally, the methodologies available to meet dialogue objectives vary widely. Given that nanotechnologies are likely to pose a wide range of issues there is no single method to draw upon. Rather, dialogue methods must be designed specifically around the objectives at hand at any point in time. Renn et al distinguish between three broad classes of citizen participation: genuine deliberative methods which allow for fair and competent debate and discussion between all parties, such as consensus conferences, citizens' juries and planning cells; traditional consultation methods, including public meetings, surveys, focus groups, and mediation, where there is little or no extended debate; and referenda, in which people do have democratic power but which are not generally deliberative in nature.

Although referenda are not a typical engagement mechanism in the UK (unlike some other European countries such as Switzerland), government and other organisations have used various forms of traditional consultation on science and technology issues in the past, while consensus conferences occurred in the UK in 1994 for plant biotechnology and in 1999 for radioactive waste management. Often dialogue processes need to be multi-stage: that is, involving different participants and methodologies at different points in time. The GM Nation—public debate on the commercialisation of agricultural biotechnology, held in the UK in 2003, is a case in point in multi-stage design. GM Nation—had three main engagement components: initial issue framing workshops with randomly selected members of the general public were followed by a series of open activities to which anybody could contribute, and finally a set of closed 'narrow-but-deep' groups, again comprising randomly selected members of the public. Experience with GM Nation—and deliberative processes elsewhere highlights several key requirements that any dialogue process involving nanotechnologies must meet and which we recommend.

First, dialogue and engagement should occur early, and before critical decisions about the technology become irreversible or 'locked in'.

Second, dialogue is not useful in and of itself, but has to be designed around specific objectives. Accordingly, clarity at an early stage about the objectives for dialogue is essential.

Third, at least some form of commitment from the sponsor to take account of outcomes is required when commissioning dialogue processes: otherwise why should organisers and participants bother?

Fourth, stakeholder and public dialogue should be properly integrated with other processes of technology assessment for nanotechnologies, as and when they occur.

For nanotechnologies, decisions will need to be sensitive to public values where significant ethical issues arise. Similarly, some of the issues associated with the convergence of nanotechnologies with other technologies, and in particular developments in bio-nanotechnologies, are likely to raise novel ethical questions in the future requiring wide public debate. This in turn suggests a requirement for periodic reflection on possible emerging ethical questions, and initiating appropriate forms of dialogue with stakeholders or the public as appropriate, as the technology matures and its tangible applications become clearer.

The arguments above suggest that for upstream issues such as nanotechnologies the quality, as well as the acceptability, of initial decisions will depend heavily upon achieving appropriate framings for risk and technology assessments at an early stage. In particular, framing needs to incorporate both social as well as technical outcomes and concerns.

Two important issues here, suggested by the research into public attitudes, would appear to be first the governance of nanotechnologies (who is to control and regulate nanotechnologies, and ensure that socially desirable goals can be identified and delivered), and second the long-term uncertainties. At a more operational level one could envisage the introduction of specific applications being accompanied by forms of stakeholder engagement on a case-by-case basis: one obvious issue would be to explore labelling requirements that people wish to see for specific classes for products, another the privacy implications of developments in sensing devices.

Unlike with some other issues of more mature technologies, nanotechnologies have so far not generated significant levels of conflict between stakeholders. However, as applications emerge and decisions are made, this situation might well change. The hope expressed in evidence submitted to the group is that methods for upstream deliberation may help society to find appropriate resolutions for potential conflicts in advance, by better anticipation of sensitive issues.

9

Future of Nanotech Revolution

The emerging technology revolution, together with the ongoing process of globalisation enabled by the information technology and continued improvements in transportation, on the one hand opens up possibilities for increased life span, economic prosperity, and quality of life, and on the other hand introduces further difficulties with privacy and ethical issues.

It has been argued that the accelerating pace of technological change may lead to a widening of the gap between rich and poor, developed and developing countries. However, increased global connectivity within the technology revolution may itself provide a vehicle for improved education and local technical capabilities that could enable poorer and less-developed regions of the world to contribute to and profit from technological advances via the "cottage industries" of the 21st century. The maturity of these trends varies. Some are already producing effects and controversy in wide public forums; others hold promise for significant effects by 2015 yet are currently less mature and are mostly discussed in advanced technology forums. The determination of "global significance" in such a foresight activity depends greatly on the level at which one examines a technology or its components.

Individual trends and applications may not rise to significance by themselves, but their collective contributions nevertheless might produce a significant trend. Even the Internet, for example, consists of a large number of applications, systems, and components-many of which might not hold up individually to a standard of global significance yet combined contribute to the overall effect. These varied contributors often come from different technical disciplines. Although multidisciplinary, such trends were grouped based on a dominant technology or a dominant concept of each trend. Note that there is always a strong element of uncertainty when projecting technological progress and implications for the future. This effort looked for potential foreseeable implications based on progress and directions in current science and technology (S&T) and did not attempt to predict or forecast exact events and timetables. Trends were gleaned from existing outlooks, testimonies, and foresights, providing collective opinions and points of view from a broad spectrum of individuals. As many of these published trends tended to be optimistic and visionary, attempts were made to provide insights on the challenges they will face, yielding a feel not only for possible implications but also for issues that may modulate their development.

The goal was to obtain a balanced perspective of current trends and directions, yielding ranges of possibilities rather than a single likely future to give a rich feel for the many possible paths that are being pursued. Such ranges of possible futures include both the optimistic and conservative extremes in technology foresights as well as ranges of optimistic and pessimistic implications of these trends. Some trends that hold promise but are unlikely to achieve global significance by 2015 are also mentioned. Although the examination of trends can yield a broad understanding of current directions, it will not include unforeseen technological breakthroughs. Unforeseen complex economic, social, ethical, and political effects on technological development will also have a major effect on what actually happens in the future. Although many computer scientists and visionary government program managers saw the potential for the Internet technology, it was practically impossible to predict whether it would become globally significant, the pace of its adoption, or its pervasive effect on social, political, and economic systems.

Genomics

By 2015, biotechnology will likely continue to improve and apply its ability to profile, copy, and manipulate the genetic basis of both plants and animal organisms, opening wide opportunities and implications for understanding existing organisms and engineering organisms with new properties. Research is even under way to create new free-living organisms, initially microbes with a minimal genome.

DNA analysis machines and chip-based systems will likely accelerate the proliferation of genetic analysis capabilities, improve drug search, and enable biological sensors. The genomes of plants and animals will likely continue to be decoded and profiled. To the extent that genes dictate function and behavior, such extensive genetic profiling could provide an ability to better diagnose human health problems, design drugs tailored for individual problems and system reactions, better predict disease predispositions, and track disease movement and development across global populations, ethnic groups, and other genetic pools. Note that a link between genes and function is generally accepted, but other factors such as the environment and phenotype play important modifying roles. Gene therapies will likely continue to be developed, although they may not mature by 2015. Genetic profiling could also have a significant effect on security, policing, and law. DNA identification may complement existing biometric technologies (e.g., retina and fingerprint identification) for granting access to secure systems (e.g., computers, secured areas, or weapons), identifying criminals through DNA left at crime scenes, and authenticating items such as fine art.

Genetic identification will likely become more commonplace tools in kidnapping, paternity, and fraud cases. Biosensors (some genetically engineered) may also aid in detecting biological warfare threats, improving food and water quality testing, continuous health monitoring, and medical laboratory analyses. Such capabilities could fundamentally change the way health services are rendered by greatly improving disease diagnosis, understanding predispositions, and improving monitoring capabilities. Such profiling may be limited by technical difficulties in decoding some genomic segments and in understanding the implications of the genetic code. Our current technology can decode nearly all of the entire human gene sequence, but errors are still an issue, since Herculean efforts are required to decode the small amount of remaining sequences. More important, although there is a strong connection between an organism's function and its genotype, we still have large gaps in understanding the intermediate steps in copying, transduction, isomer modulation, activation, immediate function, and this function's effect on larger systems in the organism.

Proteomics is the next big technological push after genomic decoding. Progress may likely rely on advances in bioinformatics, genetic code combination and sequencing, and other related information technologies. Despite current optimism, a number of technical issues and hurdles could moderate genomics progress by 2015. Incomplete understanding of sequence coding, transduction, isomer modulation, activation, and resulting functions could form technological barriers to wide engineering successes. Extensive rights to own genetic codes may slow research and ultimately the benefits of the decoding. At the other extreme, the inability to secure patents from sequencing efforts may reduce commercial funding and thus slow research and resulting benefits. In addition, investments in biotechnology have been cyclic in the past. As a result, advancements in research and development (R&D) may come in surges, especially in areas where the time to market (and thus time to return on investment) is long.

Cloning

Artificially producing genetically identical organisms through cloning will likely be significant for engineered crops, livestock, and research animals. Cloning may become the dominant mechanism for rapidly bringing engineered traits to market, for continued maintenance of these traits, and for producing identical organisms for research and production. Research will likely continue on human cloning in unregulated parts of the world with possible success by 2015, but ethical and health concerns will likely limit wide-scale cloning of humans in regulated parts of the world. Individuals or even some states may also engage in human or animal cloning, but it is unclear what they may gain through such efforts. Cloning, especially human cloning, has already generated significant controversies across the globe. Concerns include moral issues, the potential for errors and medical deficiencies of clones, questions of the ownership of good genes and genomes, and eugenics. Although some attempts at human cloning are possible by 2015, legal restrictions and public opinion may limit their extent. Fringe groups, however, may attempt human cloning in advance of legislative restrictions or may attempt cloning in unregulated countries. See, for example, the human cloning program announced by Clonaid.

Although expert opinions vary regarding the current feasibility of human cloning, at least some technical hurdles for human cloning will likely need to be addressed for safe, wide-scale use. "Attempts to clone mammals from single somatic cells are plagued by high frequencies of developmental abnormalities and lethality". Even cloned plant populations exhibit "substantial developmental and morphological irregularities". Research will need to address these abnormalities or at the very least mitigate their repercussions. Some believe, however, that human cloning may be accomplished soon if the research organisation accepts the high lethality rate for the embryo and the potential generation of developmental abnormalities.

Genetically Modified Organisms (GMOs)

Beyond profiling genetic codes and cloning exact copies of organisms and microorganisms, biotechnologists can also manipulate the genetic code of plants and animals and will likely continue efforts to engineer certain properties into life forms for various reasons. Traditional techniques for genetic manipulation (such as cross-pollination, selective breeding, and irradiation) will likely continue to be extended by direct insertion, deletion, and modification of genes through laboratory techniques. Targets include food crops, production plants, insects, and animals. Desirable properties could be genetically imparted to genetically engineered foods, potentially producing: improved taste; ultra-lean meats with reduced "bad" fats, salts, and chemicals; disease resistance; and artificially introduced nutrients (socalled "nutraceuticals").

Genetically modified organisms (GMOs) can potentially be engineered to improve their physical robustness, extend field and shelf life (e.g., the Flavr-Savr™ tomato), tolerate herbicides, grow faster, or grow in previously unproductive environments (e.g., in high-salinity soils, with less water, or in colder climates). Beyond systemic disease resistance, in vivo pesticide production has already been demonstrated (e.g., in corn) and could have a significant effect on pesticide production, application, regulation, and control with targeted release. Likewise, organisms could be engineered to produce or deliver drugs for human disease control.

Cow mammary glands might be engineered to produce pharmaceuticals and therapeutic organic compounds; other organisms could be engineered to produce or deliver therapeutics (e.g., the so-called "prescription banana"). If accepted by the population, such improved production and delivery mechanisms could extend the global production and availability of these therapeutics while providing easy oral delivery. In addition to food production, plants may be engineered to improve growth, change their constitution, or artificially produce new products. Trees will likely be engineered to optimise their growth and tailor their structure for particular applications such as lumber, wood pulp for paper, fruiting, or carbon sequestering (to reduce global warming) while reducing waste byproducts. Plants might be engineered to produce bio-polymers (plastics) for engineering applications with lower pollution and without using oil reserves. Bio-fuel plants could be tailored to minimise polluting components while producing additives needed by the consuming equipment.

Genetic engineering of microorganisms has long been accepted and used. For example, *E. coli* has been used for mass production of insulin. Engineering of bacterial properties into plants and animals for disease resistance will likely occur. Other animal manipulations could include modification of insects to impart desired behaviors, provide tagging (including GMO tagging), or prevent physical uptake properties to control pests in specific environments to improve agriculture and disease control. Research on modifying human genes has already begun and will likely continue in a search for solutions to genetically based diseases. Although slowed by recent difficulties, gene therapy research will likely continue its search for useful mechanisms to address genetic deficiencies or for modulating physical processes such as beneficial protein production or control mechanisms for cancer. Advances in genetic profiling may improve our understanding and selection of therapy techniques and provide breakthroughs with significant health benefits.

Some cloning of humans will be possible by 2015, but legal restrictions and public opinion may limit its actual extent. Controls are also likely for human modifications (e.g., clone-based eugenic modifications) for nondisease purposes. It is possible, however, that technology will enable genetic modifications for hereditary conditions (i.e., sickle cell anemia) through in vitro techniques or other mechanisms. GMOs are also having a large effect on the scientific community as an enabling technology. Not only do "knock-out" animals (animals with selected DNA sequences removed from their genome) give scientists another tool to study the effect of the removed sequence on the animal, they also enable subsequent analysis of the interaction of those functions or components with the animal's entire system. Although knock-outs are not always complete, they provide another important tool to confirm or refute hypotheses regarding complex organisms.

Issues and Implications

Extant capabilities in genomics have already created opportunities yet have generated a number of issues. As more organisms are decoded and the functional implications of genes are discovered, concerns about property and privacy rights for the sequencing will likely continue. The ability to profile an individual's DNA is already raising concerns about privacy and excessive monitoring. Examples include databases of DNA signatures for use in criminal investigations, and the potential use of genetically based health predispositions by insurance companies or employers to deny coverage or to discriminate. The latter may raise policy issues regarding acceptable and unacceptable profiling for insurance or employment. This issue is further worrisome because the exact code-tofunction mechanisms that trigger many disease predispositions are not well understood. Issues may also arise if a strong genetic basis of human physical or cognitive ability is discovered. On the positive side, understanding a person's predisposition for certain abilities (or limitations) could enable custom educational or remediation programs that will help to compensate for genetic inclinations, especially in early years when their effect can be optimised.

On the negative side, groups may use such analyses in arguments to discriminate against target populations (despite, for example, the fact that ethnic distribution variances of cognitive ability are currently believed to be wider than ethnic mean differences), aggravating social and international conflicts. Although the genetic profiles of plants have been modified for centuries using traditional techniques, questions regarding the safety of genetically modified foods have sparked international concerns in the United Kingdom and Europe, forcing a campaign by biotechnology companies to argue the safety of the technology and its applications. Some have argued that genetic engineering is actually as safe or safer than traditional combinatorial techniques such as irradiated seeds, since there often is strong supporting information concerning the function of the inserted sequences. Governments have been forced into the issue, resulting in education efforts, food labeling proposals, and heated international trade discussions between the United States and Europe on the importation of GMOs and their seedlings.

As genetic modification becomes more common, it may become more difficult to label and separate GMOs, resulting in a forcing function to resolve the issue of how far the technology should be applied and whether separate markets can be maintained in a global economy. Some have likened the anti-biotechnology movement to the anti-nuclear-power movement in scope and tactics, although the low cost and wide availability of basic genomic equipment and know-how will likely allow practically any country, small business, or even individual to participate in genetic engineering. Such wide technology availability and low entry costs could make it impossible for any movement or government to control the spread and use of genomic technology. At an extreme, successful protest pressures on big biotechnology companies together with wide technology availability could ultimately drive genomic engineering "underground" to groups outside such pressures and outside regulatory controls that help ensure safe and ethical uses. This could ironically facilitate the very problems that the anti-biotechnology movement is hoping to prevent.

Cloning and genetic modification also raise biodiversity concerns. Standardisation of crops and livestock have already increased food supply vulnerabilities to diseases that can wipe out larger areas of production. Genetic modification may increase our ability to engineer responses to these threats, but the losses may still be felt in the production year unless broad-spectrum defenses are developed. In addition to food safety, the ability to modify biological organisms holds the possibility of engineered biological weapons that circumvent current or planned countermeasures. On the other hand, genomics could aid in biological warfare defense (e.g., through improved understanding and control of biological function both in and between pathogens and target hosts as well as improved capability for engineered biosensors).

Advances in genomics, therefore, could advance a race between threat engineering and countermeasures. Thus, although genetic manipulation is likely to result in medical advances. The rate at which GMO benefits are felt in poorer countries may depend on the costs of using patented organisms, marketing demands and approaches, and the rate at which crops become ubiquitous and inseparable from unmodified strains. Consider, for example, current issues related to human immunodeficiency virus (HIV) drug development and dissemination in poorer countries. Patentability has fueled research investments, but many poorer countries with dire needs cannot afford the latest drugs and must wait for handouts or patent expiration. Globalisation, however, may fuel dissemination as multinational companies invest in food production across the globe. Also, the rewards from opening previously unproductive land for production may provide the financial incentive to pay the premium for GMOs. Furthermore, widely available genomic technology could allow academics, nonprofit small businesses, and developing countries to develop GMOs to alleviate problems in poorer regions; larger biotechnology companies will focus on markets requiring capital-intensive R&D. Finally, moral issues may play a large role in modulating the global effect of genomics trends. Some people simply believe it is improper to engineer or modify biological organisms using the new techniques. Unplanned side effects will likely support such opposition. Others are concerned with the real danger of eugenics programs or of the engineering of dangerous biological organisms.

Technologies in Drug Therapies

Beyond genetics, biotechnology will likely continue to improve therapies for preventing and treating disease and infection. New approaches might block a pathogen's ability to enter or travel in the body, leverage pathogen vulnerabilities, develop new countermeasure delivery mechanisms, or modulate or augment the immune response to recognising new pathogens. These therapies may counter the current trend of increasing resistance to extant antibiotics, reshaping the war on infections. In addition to addressing traditional viral and bacterial problems, therapies are being developed for chemical imbalances and modulation of chemical stasis. For example, antibodies are being developed that attack cocaine in the body and may be used to control addiction. Such approaches could have a significant effect on modifying the economics of the global illegal drug trade while improving conditions for users. Drug development will likely be aided by various technology trends and enablers.

Computer simulations combined with proliferating trends for molecular imaging technologies (e.g., atomic-force microscopes, mass spectroscopy, and scanning probe microscopes) may continue to improve our ability to design molecules with desired functional properties that target specific receptors, binding sites, or markers, complementing combinatorial drug search with rational drug design. Simulations of drug interactions with target biological systems could become increasing useful in understanding drug efficacy and safety.

Issues and Implications

R&D costs for drug development are currently extremely high and may even be unsustainable, with averages of approximately $600 million per drug brought to market. These costs may drive the pharmaceutical industry to invest heavily in technology advances with the goal of long-term viability of the industry. Combined with genetic profiling, drug development tailored to genotypes, chemical simulation and engineering programs, and drug testing simulations may begin to change pharmaceutical development from a broad application trial-and-error approach to custom drug development, testing, and prescription based on a deeper understanding of subpopulation response to drugs. This understanding may also rescue drugs previously rejected because of adverse reactions in small populations of clinical trials. Along with the potential for improving success rates, reducing trial costs, and opening new markets for narrowly targeted drugs, tailoring drugs to subpopulations will also have the opposite effect of reducing the size of the market for each drug. Thus, the economics of the pharmaceutical and health industries will likely change significantly if these trends come to fruition.Note that patent protection is not uniformly enforced across the globe for the pharmaceutical industry. As a result, certain regions (e.g., Asia) may continue to focus on production of non-legacy (generic) drugs, and other regions (e.g., the United States, United Kingdom, and Europe) will likely continue to pursue new drugs in addition to such low-margin pharmaceuticals.

Biomedical Engineeriging and Technology

Multidisciplinary teaming is accelerating advances and products in biomedical engineering and technology of organic and artificial tissues, organs, and materials.

Organic Tissues and Organs

Advances in tissue and organ engineering and repair are likely to result in organic and artificial replacement parts for humans. New advances in tissue regeneration and repair continue to improve our ability to resolve health problems within our bodies. The field of tissue engineering, which is barely a decade old, has already led to engineered commercial skin products for wound treatment. Growth of cartilage for repair and replacement is at the stage of clinical testing, and treatment of heart disease via growth of functional tissue by 2015 is a realistic goal. These advances will depend upon improved biocompatible (or bioabsorbable) scaffold materials, development of 3D vascularised tissues and multicellular tissues, and an improved understanding of the in vivo growth process of cellular material on such scaffolds. Research and applications of stem cell therapies will likely continue and expand, using these unspecialised human cells to augment or replace brain or body functions, organs. As the most unspecialised stem cells are found in early stage embryos or fetal tissue, an ethical debate is ensuing regarding the use of stem cells for research and therapy.

Alternatives such as the use of adult human stem cells or stem cell culturing may ultimately produce large-scale cell supplies with reduced ethical concerns. Current debates have limited U.S. government funding for stem cell research, but the potential has attracted substantial private funding. Xenotransplantations (transplantation of body parts from one species to a different species) could be improved, aided by attempts tc genetically modify donor tissue and organ antibodies, complements, and regulatory proteins to reduce or eliminate rejection. Baboons or pigs, for example, may be genetically modified and cloned to produce organs for human transplant, although large-scale success may not occur by 2015. Beyond rejection, the significance of xenotransplants is likely to be modulated by concerns that diseases such as retro viruses might jump from animals to people as a result of the transplantation techniques. Ethical (e.g., animal rights) and moral concerns as well as possible patenting issues may also result in regulations and limitations on xenotransplants, limiting their significance.

Artificial Materials, Organs, and Bionics

In addition to organic structures, advances are likely to continue in engineering artificial tissues and organs for humans. Multi-functional materials are being developed that provide both structure and function or that have different properties on different sides, enabling new applications and capabilities. For example, polymers with a hydrophilic shell around a hydrophobic core (biomimetic of micelles) can be used for timed release of hydrophobic drug molecules, as carriers for gene therapy or immobilised enzymes, or as artificial tissues. Sterically stabilised polymers could also be used for drug delivery.

Other materials are being developed for various biomedical applications. Fluorinated colloids, for example, are being developed that take advantage of the high electronegativity of fluorine to enhance in vivo oxygen transport (as a blood substitute during surgery) and for drug delivery. Hydrogels with controlled swelling behavior are being developed for drug delivery or as templates to attach growth materials for tissue engineering. Ceramics such as bioactive calcia-phosphate-silica glasses (gel-glasses), hydroxyapetite, and calcium phosphates can serve as templates for bone growth and regeneration. Bioactive polymers (e.g., polypeptides) can be applied as meshes, sponges, foams, or hydrogels to stimulate tissue growth.

Coatings and surface treatments are being developed to increase biocompatibility of implanted materials (for example, to overcome the lack of endothelial cells in artificial blood vessels and reduce thrombosis). Blood substitutes may change the blood storage and retrieval systems while improving safety from blood-borne infections. New manufacturing techniques and information technology are also enabling the production of biomedical structures with custom sizing and shape. For example, it may become commonplace to manufacture custom ceramic replacement bones for injured hands, feet, and skull parts by combining computer tomography and "rapid prototyping" to reverse engineer new bones layer by layer.

Beyond structures and organs, neural and sensor prosthetics could begin to become significant by 2015. Retinas and cochlear implants, bypasses of spinal and other nerve damage, and other artificial communications and stimulations may improve and become more commonplace and affordable, eliminating many occurrences of blindness and deafness. This could eliminate or reduce the effect of serious handicaps and change society's response from accommodation to remediation.

Biomimetics and Applied Biology

Recent techniques such as functional brain imaging and knock-out animals are revolutionising our endeavors to understand human and animal intelligence and capabilities. These efforts should, by 2015, make significant inroads in improving our understanding of phenomena such as false memories, attention, recognition, and information processing, with implications for better understanding people and designing and interfacing artificial systems such as autonomous robots and information systems. Neuromorphic engineering has already produced novel control algorithms, vision chips, head-eye systems, and biomimetic autonomous robots. Although not likely to produce systems with wide intelligence or capabilities similar to those of higher organisms, this trend may produce systems by 2015 that can robustly perform useful functions such as vacuuming a house, detecting mines, or conducting autonomous search.

Biotechnology and materials advances are likely to continue producing revolutionary surgical procedures and systems that will significantly reduce hospital stays and cost and increase effectiveness. New surgical tools and techniques and new materials and designs for vesicle and tissue support will likely continue to reduce surgical invasiveness and offer new solutions to medical problems. Techniques such as angioplasty may continue to eliminate whole classes of surgeries; others such as laser perforations of heart tissue could promote regeneration and healing. Advances in laser surgery could refine techniques and improve human capability, especially as costs are reduced and experience spreads. Hybrid imaging techniques will likely improve diagnosis, guide human and robotic surgery, and aid in basic understanding of body and brain function.

Finally, collaborative information technology (e.g., "telemedicine") will likely extend specialised medical care to remote areas and aid in the global dissemination of medical quality and new advances.

Issues and Implications

By 2015, one can envision: effective localised, targeted, and controlled drug delivery systems; long-lived implants and prosthetics; and artificial skin, bone, and perhaps heart muscle or even nerve tissue. A host of social, political, and ethical issues such as those discussed above will likely accompany these developments. Biomedical advances are already increasing human life span in countries where they are applied. New advances by 2015 are likely to continue this trend, accentuating issues such as shifts in population age demographics, financial support for retired persons, and increased health care costs for individuals. Advances, however, may improve not only life expectancy but productivity and utility of these individuals, offsetting or even overcoming the resulting issues.

Many costly and specialised medical techniques are likely to initially benefit citizens who can afford better medical care; wider global effects may occur later as a result of traditional trickle-down effects in medicine. Some technologies may have the opposite trend where low-cost technologies may enable cost-effective consulting with specialists regardless of location. However, access to technology may greatly mediate this dispersal mechanism and may place additional demands on technology upgrades and education. Countries that remain behind in terms of technological infrastructures may miss many of these benefits. Theological debates have also raised concerns about the definition of what constitutes a human being, since animals are being modified to produce human organs for later xenotransplantation in humans. Improved understanding of human intelligence and cognitive function could have broader legal and social effects. For example, an understanding of false memories and how they are created could have an effect on legal liabilities and courtroom testimony. Understanding innate personal capabilities and job performance requirements could help us determine who would make better fighter pilots, who has an edge in analysing complex images, and what types of improved training could improve people's capabilities to meet the special demands of their chosen careers.

Ethical concerns could arise concerning discrimination against people who lack certain innate skills, requiring objective and careful measures for hiring and promotion. Eventually, neural and sensory implants could radically change the way people sense, perceive, and interact with natural and artificial envi-ronments. Ultimately, these new capabilities could create new jobs and functions for people in these environments. Such innovations may first develop for individuals with particularly challenging and critical functions (e.g., soldiers, pilots, and controllers), but innovations may first develop in other quarters (e.g., for entertainment or business functions), given recent trends. Initial research indicates the feasibility of such implants and interactions, but it is unclear whether R&D and investments will accelerate enough to realise even such early applications by 2015. Current trends have concentrated on medical prosthetics where research prototypes are already appearing so it appears likely that globally significant systems will appear in this domain first.

Materials Engineering

New materials can often be critical enabling drivers for new systems and applications with significant effects. However, it may not be obvious how enabling materials affect more observable trends and applications. A common process model from materials engineering can help to show how materials appear likely to break previous barriers in the process that ultimately results in applications with potential global benefits. Developments in materials science and engineering result from interdisciplinary materials research. This development can be conveniently represented by the schematic description of the materials engineering process from concept to product/application (Figure 1).

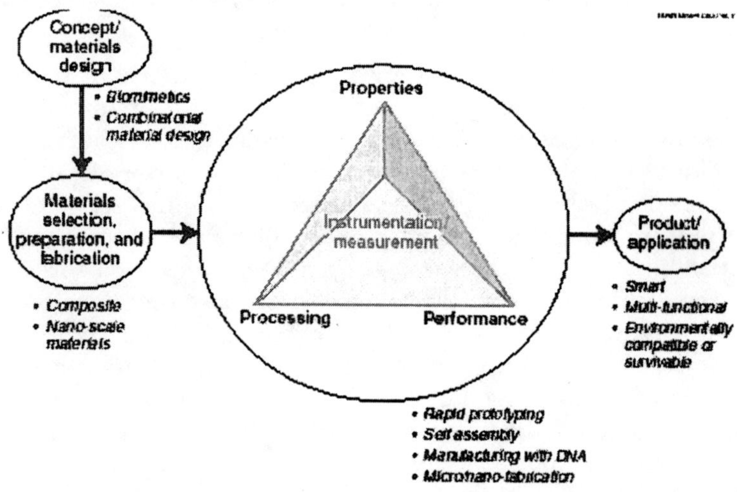

Figure 1. The General Materials Engineering Process

Concept/Materials Design

Biomimetics is the design of systems, materials, and their functionality to mimic nature. Current examples include layering of materials to achieve the hardness of an abalone shell or trying to understand why spider silk is stronger than steel. Combinatorial materials design uses computing power to screen many different materials possibilities to optimise properties for specific applications (e.g., catalysts, drugs, optical materials).

Selection, Preparation, and Fabrication

Composites are combinations of metals, ceramics, polymers, and biological materials that allow multi-functional behavior. One common practice is reinforcing polymers or ceramics with ceramic fibers to increase strength while retaining light weight and avoiding the brittleness of the monolithic ceramic. Materials used in the body often combine biological and structural functions (e.g., the encapsulation of drugs). Nanoscale materials, i.e., materials with properties that can be controlled at submicrometer ($<10^{-6}$ m) or nanometer (10^{-9} m) level, are an increasingly active area of research because properties in these size regimes are often fundamentally different from those of ordinary materials. Examples include carbon nanotubes, quantum dots, and biological molecules. These materials can be prepared either by purification methods or by tailored fabrication methods.

These areas are inextricably linked to each other: Processing determines properties that in turn determine performance. Moreover, the sensitivity of instrumentation and measurement capability is often the enabling factor in optimising processing, for example, as for nanotechnology and microelectromechanical systems (MEMS).

Rapid prototyping is the capability to combine computer-assisted design and manufacturing with rapid fabrication methods that allow inexpensive part production (as compared to the cost of a conventional production line). Rapid prototyping enables a company to test several different inexpensive prototypes before committing infrastructure investments to an approach. Combined with manufacturing system improvements to allow flexibility of approach and machinery, rapid prototyping can lead to an agile manufacturing capability.

Alternatively, the company can use its virtual capability to design and then outsource product manufacturing, thus offloading capital investment and risk. This capability is synergistic with the information technology revolution in the sense that it is a further factor in globalising manufacturing capability and enabling organisations with less capital to have a significant technological effect.

For the Department of Defense (DoD), it could reduce or eliminate requirements for warehousing large amounts of spares and, for example, could enable the Air Force to "fly before they buy." Self-assembly refers to the use in materials processing or fabrication of the tendency of some materials to organise themselves into ordered arrays (e.g., colloidal suspensions). This provides a means to achieve structured materials "from the bottom up" as opposed to using manufacturing or fabrication methods such as lithography, which is limited by the measurement and instrumentation capabilities of the day.

Organic polymers have been tagged with dye molecules to form arrays with lattice spacing in the visible optical wavelength range and that can be changed through chemical means. This provides a material that fluoresces and changes color to indicate the presence of chemical species. Manufacturing with DNA might represent the ultimate biomimetic manufacturing scheme. It consists of "functionalising small inorganic building blocks with DNA and then using the molecular recognition processes associated with DNA to guide the assembly of those particles or building blocks into extended structures". Using this approach, Mirkin and colleagues demonstrated a highly selective and sensitive DNA-based chemical assay method using 13 nm diameter gold parti-cles with attached DNA sequences. This approach is compatible with the commonly used polymerase chain reaction (PCR) method of amplification of the amount of the target substance. Micro- and nano-fabrication methods include, lithography of coupled micro- or nano-scale devices on the same semiconductor or biological material. It is important to note the crucial role played in the development of these techniques by the parallel development of instrumentation and measurement devices such as the Atomic Force Microscope (AFM) and the various Scanning Probe Microscopes (SPMs).

The trends described above will likely work in concert to provide materials engineers with the capability to design and produce advanced materials that will be:

— *Smart*—Reactive materials combining sensors and actuators, perhaps together with computers, to enable response to environmental conditions and changes thereof. An example might be robots that mimic insects or birds for applications such as space exploration, hazardous materials location and treatment, and unmanned aerial vehicles (UAVs).

— *Multi-functional*—MEMS and the "lab-on-a-chip" are excellent examples of systems that combine several functions. Another example is a drug delivery system using a hydrogel with hydrophilic exterior and hydrophobic interior. Consider also aircraft skins fabricated from radar-absorbing materials that incorporate avionic links and the ability to modify shape in response to airflow.

— *Environmentally compatible or survivable*—The development of composite materials and the ability to tailor materials at the atomic level will likely provide opportunities to make materials more compatible with the environments in which they will be used. Examples might include prosthetic devices that serve as templates for the growth of natural tissue and structural materials that strengthen during service (e.g., through temperature or stress-induced phase changes).

Technology for Smart Materials

Several different types of materials exhibit sensing and actuation capabilities, including ferroelectrics (exhibiting strain in response to a electric field), shape-memory alloys (exhibiting phase transition-driven shape change in response to temperature change), and magnetostrictive materials (exhibiting strain in response to a magnetic field).

These effects also work in reverse, so that these materials, separately or together, can be used to combine sensing and actuation in response to environmental conditions. They are currently in widespread use in applications from ink-jet printers to magnetic disk drives to anti-coagulant devices. An important class of smart materials is composites based upon lead zirconate titanate (PZT) and related ferroelectric materials that allow increased sensitivity, multiple frequency response, and variable frequency.

Applications include sensors and actuators that can change their frequency either to match a signal or to encode a signal. Ferroelectrics are already in use as nonvolatile memory elements for smart cards and as active elements in smart skis that change shape in response to stress. Another important class of materials is smart polymers (e.g., ionic gels that deform in response to electric fields). Such electro-active polymers have already been used to make "artificial muscles".

Currently available materials have limited mechanical power, but this is an active research area with potential applications to robots for space exploration, hazardous duty of various types, and surveillance. Hydrogels that swell and shrink in response to changes in pH or temperature are another possibility; these hydrogels could be used to deliver encapsulated drugs in response to changes in body chemistry (e.g., insulin delivery based upon glucose concentration). Another variation on this trend for controlled release of drugs is materials with hydrophilic exterior and hydrophobic interior.

Issues and Implications

A world with pervasive, networked sensors and actuators (e.g., on and part of walls, clothing, appliances, vehicles, and the environment) promises to improve, optimise, and customise the capability of systems and devices through availability of information and more direct actuation. Continuously available communication capability, ability to catalog and locate tagged personal items, and coordination of support functions have been espoused as benefits that may begin to be realised by 2015.

The continued development of small, low-profile biometric sensors, coupled with research on voice, handwriting, and fingerprint recognition, could provide effective personal security systems. These could be used for identification by police/military and also in business, personal, and leisure applications. Combined with today's information technologies, such uses could help resolve nagging security and privacy concerns while enabling other applications such as improved handgun safety (through owner identification locks) and vehicle theft control. Other potential applications of smart materials that would be enabled by 2015 include: clothes that respond to weather, interface with information systems, monitor vital signs, deliver medicines, and automatically protect wounds; airfoils that respond to airflow; buildings that adjust to the weather; bridges and roads that sense and repair cracks; kitchens that cook with wireless instructions; virtual reality telephones and entertainment centers; and personal medical diagnostics.

The level of development and integration of these technologies into everyday life will probably depend more on consumer attitudes than on technical developments. In addition to the surveillance and identification functions mentioned under smart materials above, developments in robotics may provide new and more sensitive capabilities for detecting and destroying explosives and contraband materials and for operating in hazardous environments. Increases in materials performance, both for power sources and for sensing and actuation, as well as integration of these functions with computing power, could enable these applications. Such trend potentials are not without issues. Pervasive sensory information and access to collected data raise significant privacy concerns. Also, the pace of development will likely depend on investment levels and market drivers. In many cases the immediate benefits and cost savings from smart material applications will continue to drive development, but more exotic materials research may depend on public commitment to research and belief in investing in longer-term rewards.

Technology for Rapid Prototyping

This manufacturing approach integrates computer-aided design (CAD) with rapid forming techniques to rapidly create a prototype that can be used to visualise or test the part before making the investment in tooling required for a production run. Originally, the prototypes were made of plastic or ceramic materials and were not functional models, but now the capability exists to make a functional part, e.g., out of titanium.

Building Using Materials

Research on composite materials, waste management, and recycling has reached the stage where it is now feasible to construct buildings using materials fabricated from significant amounts of indigenous waste or recycled material content. These approaches are finding an increasing number of cost-effective applications, especially in developing countries. Examples include the Petronas

Twin Towers in Kuala Lumpur, Malaysia. These towers are the tallest buildings on earth and are made with reinforced concrete rather than steel. A roofing material used in India is made of natural fiber and agro-industrial waste. Prefabricated composite materials for home construction have also been developed in the United States, and a firm in the Netherlands is developing a potentially ubiquitous, inexpensive housing approach targeted for developing countries that uses spray-forming over an inflatable air shell.

Transportation

An important trend in transportation is the development of lightweight materials for automobiles that increase energy efficiency while reducing emissions. Here the key issue is the strength-to-weight ratio versus cost. Advanced composites with polymer, metal, or ceramic matrix and ceramic reinforcement are already in use in space systems and aircraft.

These composites are too expensive for automobile applications, so aluminum alloys are being developed and introduced in cars such as the Honda Insight, the Audi A8 and AL2, and the GM EV1. Although innovation in both design and manufacturing is needed before such all-aluminum structures can become widespread, aluminum content in luxury cars and light trucks has increased in recent years. Polymer matrix, carbon-fiber (C-fiber) reinforced composites could enable high mileage cars, but C-fiber is currently several times more expensive than steel. Research sponsored by the Department of Energy (DOE) at Oak Ridge National Laboratory is working to develop cheap C-fibers which could have wider application and effect. Spurred by California's regulations concerning ultra-low-emission vehicles, both Honda and Toyota have introduced gasoline-electric hybrid vehicles.

The U.S. government and industry consortium called Partnership for a New Generation of Vehicles (PNGV) has demonstrated prototype hybrid vehicles that use both diesel/electric and diesel/fuel-cell power plants and has established 2008 as the goal for a production vehicle. These vehicles use currently available materials, but the cost reduction issues described above will be critical in bringing production costs to levels that will allow significant market penetration.

Energy System

If the ready availability of oil continues, it may be difficult for technology trends to be much of a driving force in global energy between now and 2015. Key questions have to do with continued oil imports, continued use of coal, sources of natural gas, and the fate of nuclear power. Nevertheless, technology may have significant effects in some areas. Along with investments in solar energy, current investments in battery technology and fuel cells could enable continued trends in more portable devices and systems while extending operating times.

Developments in materials science and engineering may enable the energy systems of 2015 to be more distributed with a greater capability for energy storage, as well as energy system command, control, and communication. High-temperature superconducting cables, transformers, and storage devices could begin to increase energy transmission and distribution capabilities and power quality in this time frame. The continued development of renewable energy could be enhanced by the combination of cheap, lightweight, recyclable materials to provide cost-effective energy for developing countries without existing, well-developed energy infrastructures as well as for remote locations.

Significant changes in developed countries, however, may be driven more by existing social, political, and business forces, since the fuel mix of 2015 will still be strongly based on fossil fuels. Environmental concerns such as global warming and pollution might shift this direction, but it would likely require long-term economic problems (e.g., a prolonged rise in the price of oil) or distribution problems to drive advances in renewable energy development.

New Materials

Materials research may provide improvements in properties by 2015 in a number of additional areas, leading to significant effects. SiC, GaN, and other wide band gap semiconductors are being investigated as materials for high-power electronics. Functionally graded materials (i.e., materials whose properties change gradually from one end to the other) can form useful interlayers between mechanically, thermally, or electrically diverse components. Anodes, cathodes, and electrolytes with higher capacity and longer lifetime are being developed for improved batteries and fuel cells.

High-temperature superconductors discovered in 1986 can currently operate at liquid nitrogen temperatures. Prototype devices such as electrical transmission cables, transformers, storage devices, motors, and fault current limiters have now been built and demonstrated. Niche application on electric utility systems should begin by 2015 (e.g., replacement of underground cables in cities and replacement of older substation transformers). Nonlinear optical materials such as doped $LiNbO_3$ are being investigated for ultraviolet lasers (e.g., to enable finer lithography). Efforts are under way to increase damage threshold and conversion efficiency, minimise divergence, and tailor the absorption edge. Hard materials such as nanocrystalline coatings and diamonds are being developed for applications such as computer disk drives and drill bits for oil and gas exploration, respectively. High-temperature materials such as ductile intermetallics and ceramic matrix composites are being developed for aerospace applications and for high-efficiency energy and petrochemical conversion systems.

Nanomaterials

This area combines nanotechnology and many applications of nanostructured materials. One important research area is the formation of semiconductor "quantum dots" (i.e., several nanometer-size, faceted crystals) by injecting precursor materials conventionally used for chemical-vapor deposition of semiconductors into a hot liquid surfactant. This "quantum dot" is in reality a macromolecule because it is coated with a monolayer of the surfactant, preventing agglomeration.

These materials photoluminesce at different frequencies depending upon their size, allowing optical multiplexing in biological labeling. Another important class of nanomaterials is nanotubes (the open cylindrical sisters of fullerenes). Possible applications are field-emission displays, nanoscale wires for batteries, storage of Li or H_2, and thermal management (heat pipes or insulation-the latter taking advantage of the anisotropy of thermal conductivity along and perpendicular to the tube axis).

Another possibility is to use nanotubes (or fibers built from them) as reinforcement for composite materials. Presumably because of the nature of the bonding, it is predicted that nanotube-based material could be 50 to 100 times stronger than steel at one-sixth of the weight if current technical barriers can be overcome.

Nanoscale structures with desirable mechanical and other properties may also be obtained through processing. Examples include strengthening of alloys with nanoscale grain structure, increased ductility of metals with multi-phase nanoscale microstructure, and increased flame retardancy of plastic nanocomposites. Much has been made of the trend toward producing devices with ever-decreasing scale. Many people have projected that nanometer-scale devices will continue this trend, bringing it to unprecedented levels. This includes scale reduction not only in microelectronics but also in fields such as MEMS and quantum-switch-based computing in the shorter term.

Nanofabricated Chips. SEMATECH-the leading industry group in the semiconductor manufacturing business-is calling for the development of nanoscale semiconductors in their latest International Technology Roadmap for Semiconductors (ITRS). The roadmap calls for a 35 nm gate length in 2015 with a total number of functions in high-volume production microprocessors of around 4.3 billion. For low-volume, high-performance processors, the number of functions may approach 20 billion.

Corresponding memory chips (DRAMs) are targeted to hold around 64 gigabytes. These roadmap targets would continue the exponential trend in processing power, fueling advances in information technology. Although a number of engineering challenges exist (such as lithography, interconnects, and defect management), obstacles to achieve at least this level of performance do not seem insurmountable. Given unforeseen shortfalls in the economic production of these chips (e.g., because of very high manufacturing costs or unacceptably large numbers of manufacturing defects), several alternatives seem possible. Defect-tolerant computer architectures such as those prototyped on a small scale by Hewlett-Packard offer one alternative. These alternative methods provide some level of additional robustness to the performance goals set by the ITRS. However, in the years following 2015, additional difficulties will likely be encountered, some of which may pose serious challenges to traditional semiconductor manufacturing techniques.

In particular, limits to the degree that interconnections or "wires" between transistors may be scaled could in turn limit the effective computation speed of devices because of materials properties and compatibility, despite incremental present-day advances in these areas. Thermal dissipation in chips with extremely high device densities will also pose a serious challenge. This issue is not so much a fundamental limitation as it is an economic consideration, in that heat dissipation mechanisms and cooling technology may be required that add to total system cost, thereby adversely affecting marginal cost per computational function for these devices.

Quantum-switch-based computing. One potential long-term solution for overcoming obstacles to increased computational power is computing based on devices that take advantage of various quantum effects. The core innovation in this work is the use of quantum effects, such as spin polarisation of electrons, to determine the state of individual switches. This is in contrast to more traditional microelectronics, which are based on macroscopic properties of large numbers of electrons, taking advantage of materials properties of semiconductors.

Various concepts of quantum computers are attractive because of their massive parallelism in computation, but they are not anticipated to have significant effects by 2015. These concepts are qualitatively different from those employed in traditional computers and will hence require new computer architectures. The types of computations (and hence applications) that can be quickly performed using these computers are not the same as those readily addressed by today's digital computer.

Several workers in the area have devised algorithms for problems that are very computationally intensive (and thus time-consuming) for existing digital computers, which could be made much faster using the physics of quantum computers. Examples of these problems include factoring large numbers, searching large databases, pattern matching, and simulation of molecular and quantum phenomena.

A preliminary survey of work in this area indicates that quantum switches are unlikely to overcome major technical obstacles, such as error correction, de-coherence and signal input/output, within the next 15 years. If this were indeed the case, quantum-switch-based computing does not appear to be competitive with traditional digital electronic computers within the 2015 timeframe.

Many of the same manufacturing and architectural challenges discussed above regarding quantum computing also hold true for molecular electronics. Molecular electronic devices could operate as logic switches through chemical means, using synthesised organic compounds. These devices can be assembled chemically in large numbers and organised to form a computer.

The main advantage of this approach is significantly lower power consumption by individual devices. Several approaches for such devices have been devised, and experiments have shown evidence of switching behavior for individual devices. Several research groups have proposed interconnection between devices using carbon nanotubes, which provide high conductivity using single molecular strands of carbon.

Progress has been made toward raising the operating temperature of these switches to nearly room temperature, making the switching process reversible, and increasing the overall amount of current that can be switched using these devices. Several major outstanding issues remain with respect to molecular electronics. One issue is that molecular memories must be able to maintain their state, just as in a digital electronic computer.

Also, given that the manufacturing and assembly process for these devices will lead to device defects, a defect-tolerant computer architecture needs to be developed. Fabricating reliable interconnects between devices using carbon nanotubes is an additional challenge. A significant amount of work is ongoing in each of these areas. Even though experimental progress to date in this area has been substantial, it seems unlikely that molecular computers could be developed within the next 15 years that would be relatively attractive compared with conventional electronic computers.

Issues and Implications

Examining the potential for developing qualitatively different computational capabilities from different technology bases is a challenging exercise. The history of computing over the last 50 years has seen one major shift in technology base (from vacuum tubes to semiconductor transistors), with a corresponding shift not just in computational power but also in attitudes about the value of computers. Ideas of computers as simple machines for computation gave way to the use of computers for personal productivity with the advent of the microprocessor.

As the power of these microprocessors has grown exponentially, they have also been seen more recently as a vehicle for new media and socialisation.The ramifications of future computing technologies will be determined principally by two factors: the conception, development, and adoption of new applications that require significantly more computational power; and the ability of technology to address these demands. New applications are always difficult to anticipate, but it is less challenging to foresee the likely consequences for diffusion of this technology.

Past experience with personal computers and telecommunications has shown that these technologies diffuse more rapidly in the developed world than in the developing world. It is difficult to foresee an increase in the political or ethical barriers to computing technologies beyond those seen today, and these are rapidly vanishing. On the remaining question of technology development, the odds-on favorite for the next 15 years remains traditional digital electronic computers based on semiconductor technology. Given the virtual certainty of continued progress in this area, it is hard to imagine a scenario in which a competing technology could offer a significant performance advantage at a competitive price.

But the longer-term, traditionally elusive question in the period after 2015 is: How long will traditional silicon computing last? And when, if ever, will a competing technology become available and attractive? If an alternative computing technology becomes sufficiently attractive, the economic effects of technology substitution on the current semiconductor industry and adjacent industries must be considered.

Major industry players may be faced with a choice between cannibalising their existing market opportunities in favor of these new, future technologies, competing head-on with new players, or simply acquiring them.

Most important, given the very different architectural approaches of these technologies and the classes of problems for which they are best suited, what will be the effect on future applications? The promises of nanotechnology may indeed become a reality in the period after 2015, but it will face these competitive challenges before its significance becomes global.

Microsystems and MEMs

MEM is less an application area in itself than a manufacturing or fabrication technique that enables other application areas. Many authors use MEMS as shorthand to imply a number of particular application areas. As it is used here, MEMS is a "topdown" fabrication technology that is especially useful for integrating mechanical and electrical systems together on the same chip. It is grouped in the category of integrated microsystems because these same MEMS techniques can be extended in the future to also help integrate biological and chemical components on the same chip, as discussed below. Thus far, MEMS techniques have been used to make some functional commercial devices such as sensors and single-chip measurement devices. Many researchers have used MEMS technologies as analytical tools in other areas of nanotechnology such as the ones discussed here.

Smart Systems-on-a-Chip

Simple electro-optical and chemical sensor components have already been successfully integrated onto logic and memory chip designs in research and development labs. Likewise, radio frequency component integration in wireless devices is already being produced in mass quantities. Some companies have products capable of doing elementary DNA testing. The 1999 ITRS predicts the introduction of chemical sensor components with logic in commercial designs by 2002, with electro-optical component integration by 2004, and biological systems integration by 2006. Given these predictions, there is clearly time for relatively complex integrated systems and applications to develop within the 2015 timeline. These advances could enable many applications where increased integrated functionality can become ubiquitous as a result of lower costs and micro-packaging.

Instrumentation and measurement technology

Instrumentation and measurement technologies are some of the most promising areas for near-term advancements and enabling effects. As optical, fluidic, chemical, and biological components can be integrated with electronic logic and memory components on the same chip at marginal cost, drug discovery, genetics research, chemical assays, and chemical synthesis are all likely to be substantially affected by these advances by 2015. Some of the first applications of nanoscale (and microscale) instruments were as basic sensors for acceleration (such as those used in airbags), pressure, etc.

Small, microscale, special-purpose optical and chemical sensors have been used for some time in sophisticated laboratory equipment, along with microprocessors for signal processing and computation. Already, companies have produced products that allow for basic DNA analysis, and that can assist in drug discovery. As these sensors become more sophisticated and more integrated with computational capability (with the aid of systems-on-a-chip), their utility should grow tremendously, especially in the biomedical arena.

Issues and Implications

There are several advantages of nanotechnology for integrated systems in general (and instrumentation and measurement systems as a subset of these). First, existing semiconductor technologies will likely allow the volume manufacture of integrated smart systems that can be produced at low enough cost to be considered disposable. Second, the massive parallelism afforded by this same technology allows for the rapid analysis (with integrated computation) of very complex samples (such as DNA), the processing of large numbers of samples, and the recognition of large numbers of agents (e.g., infectious agents and toxins). Devices with these properties are already of tremendous utility in the biomedical arena for drug testing, chemical assays, etc. In addition, they will likely find utility in a variety of industrial applications.

Integrated micro/nanosystems are already starting to affect applications where miniaturisation of components, subsystems, and even complete systems is significantly reducing device size, power, and consumables while introducing new capabilities. The next five to ten years will likely see the integration of computational capabilities with biological, chemical, and optical components in systems-on-a-chip.

At the same time, advances in biotechnology should drive applications for drug discovery and genomics, as well as the basic understanding of many other phenomena. Advances in biomaterials will likely produce biologically compatible packaging, capable of isolating substances from the body in a time-controlled fashion (e.g., for drug delivery).

The confluence of these capabilities could allow for continued development of microscale and nanoscale systems that could continue to be introduced into the body to perform basic diagnostic functions in a minimally invasive way, providing new abilities to remedy health problems. Other possible applications include: pervasive, self-moving sensor systems; nanoscrubbers and nanocatalysts; even inexpensive, networked "nanosatellites."

So-called "nanosatellites" are targeting order-of-magnitude reductions in both size and mass (e.g., down to 10 kg) by reducing major system components using integrated microsystems. If successful, this could economise current missions and approaches (e.g., communication, remote sensing, global positioning, and scientific study) while enabling new missions (e.g., military tactical space support and logistics, distributed sparse aperture radar, and new scientific studies). In addition, advances could empower the proliferation of currently controlled processing capabilities (e.g., nuclear isotope separation) with associated threats to international security. Progress will likely depend on investment levels as well as continued S&T development and progress.

Manufacturing Nanorobots

A number of experts have put forth the concept of

molecular manufacturing where objects are assembled atom by atom (or molecule by molecule). Bottom-up molecular manufacturing differs from microtechnology and MEMS in that the latter employ top-down approaches using bulk materials using macroscopic fabrication techniques. To realise molecular manufacturing, a number of technical accomplishments are necessary. First, suitable molecular building blocks must be found. These building blocks must be physically durable, chemically stable, easily manipulated, and (to a certain extent) functionally versatile. Several workers in the field have suggested the use of carbon-based diamond-like structures as building blocks for nano-mechanical devices, such as gears, pivots, and rotors.

Other molecules could also be used to build structures, and to provide other integrated capabilities, such as chemically reactive structures. Much additional work in the area of modeling and synthesis of appropriate molecular structures is needed, and a number of groups are working to this end. Dresselhaus and others are fabricating suitable molecular building blocks for these structures. The second major area for development is in the ability to assemble complex structures based on a particular design.

A number of researchers have been working on different approaches to this issue. Different techniques for physical placements are under development. One approach by Quate, MacDonald, and Eigler uses atomicforce or molecular microscopes with very small nanoprobes to move atoms or molecules around with the aid of physical or chemical forces.

An alternative ap-proach by Prentiss uses lasers to place molecules in a desired location. Chemical assembly techniques are being addressed by a number of groups, including Whiteside's approach to building structures one molecular layer at a time. A third major area for development within molecular manufacturing is systems design and engineering. Extremely complex molecular systems at the macro scale will require substantial subsystem design, overall system design, and systems integration, much like complex manufactured systems of the present day.

Although the design issues are likely to be largely separable at a subsystems level, the amount of computation required for design and validation is likely to be quite substantial. Performing checks on engineering constraints, such as defect tolerance, physical integrity, and chemical stability, will be required as well. Some workers in the area have outlined a potential path for the evolution of molecular manufacturing capability, which is broken down by overall size, type of fabrication technology, system complexity, component materials used, etc.

Some versions of this concept foresee the use of massively parallel nanorobots or scanning nanoprobes to assemble structures physically. Other, more advanced concepts incorporate chemical principles and use simple chemical feedstocks to achieve much larger devices on the order of 10^8 to 10^9 molecular parts. Ostensibly, as each of these techniques matures (or fails to develop), more systems and engineering-level work must be done before applications can be realised on a significant scale. Although molecular manufacturing holds the promise of significant global changes (such as retraining large numbers of manufacturing workforces, opportunities for new regions to vie for dominance in a new manufacturing paradigm, or a shift to countries that do not have legacy manufacturing infrastructures), it remains the least concrete of the technologies discussed here.

Significant progress has been made, however, in the development of component technologies within the first regime of molecular manufacturing, where objects might be constructed from simple molecules and manufactured in a short amount of time via parallel atomic force microprobes or from simple self-assembled structures. Although the building blocks for these systems currently exist only in isolation at the research stage, it is certainly reasonable to expect that an integrated capability could be developed over the next 15 years.

Such a system could be able to assemble structures with between 100 and 10,000 components and total dimensions of perhaps tens of microns. A series of important breakthroughs could certainly cause progress in this area to develop much more rapidly, but it seems very unlikely that macro-scale objects could be constructed using molecular manufacturing within the 2015 timeframe.

Trends and Effects

The progress and effect of these trends will be modulated by enablers and barriers. Furthermore, these trends could have various effects on the world. Investments and genome decoding are fueling the ability to modify and engineer organisms to provide needed capabilities, but social concerns are already affecting the generation and use of GM foods, especially between the United States and European Union. In an optimistic 2015, GM foods will be widespread, resulting in significant benefits for food quality, global production, and the environment. Policy controls or lack of investments might moderate the production and use of GM foods, leading to increased reliance on traditional mechanisms for food productivity increases and pest control. Investments and commitment to research are prime enablers, but limited funding, limited labor, failing interests, or lack of public acceptance of highly monitored environments could modulate growth and application.

In an optimistic 2015, smart materials could be used in a wide array of novel applications. Barriers, however, could slow the development and application of smart materials to, say, advanced sensors with integrated actuator capabilities. Current high-visibility investments and technology breakthroughs will be needed to realise the full potential of nanotechnology, but research and development costs, applicability, complexity, accessibility, and even social acceptance could slow its growth. The optimistic future state is perhaps best exemplified by the vision of pervasive nanotechnology involving molecular manufacturing of a host of nanosystems with revolutionary capabilities; moreover, nanomanufacturing would take place on a global scale, giving developing countries the opportunity to invest in and participate in the revolution.

From a more pragmatic view, lack of technological breakthroughs might limit the results by 2015 to an evolutionary path where the current trend to smaller, faster, and cheaper systems continues through nano-level advances in semiconductor production to continue Moore's Law. The relationships emphasise that well-know technologies such as information technology and biotechnology actually rely on less-known enabling technologies for some of their progress. Although these facilitative relationships impart dependency on other technologies, the combined effect will accelerate the capability and promise of technology as long as key enablers can be maintained.

A number of meta-trends can be observed by reviewing the technology trends discussed above and the discussions in the open literature. These meta-trends include the increasingly multidisciplinary nature of technology, the accelerating pace of change and concerns, increasing educational demands, increased life spans, the potential for reduced privacy, continued globalisation, and the effects of international competition on technology development.

Many technology trends have been enabled by the contributions of two or more intersecting technologies. Consider, for example, MEMS-based molecular diagnostics, biomaterials, biological-based computing, and biomimetic robotics. Various technologies have combined in the past to enable applications, but there has been an increase in multidisciplinary teaming to examine system challenges and envision approaches in a unified way rather than through a hierarchical relationship.

Materials scientists, for example, are working increasingly with computer scientists and application engineers to develop biomedical materials for artificial tissues or to develop reactive materials to facilitate active system control surfaces. Materials are also being developed and adapted as embedded sensors and actuators for smart structures.

Materials provide function, and the emergence of nanotechnology has enabled construction at a scale that integrates function with processing (smart materials). The combination of materials technology with biology and the life sciences has at the same time provided knowledge and materials obtained from living organisms to enable a further integration with pervasive effects from design (biomimetics) to end product (bionics). Note that these examples show different combinations of technologies resulting in different advances; the entire set of technologies provides a rich mix of contributions to the overall technology revolution.

In addition to their technology and artifacts, different fields also tend to produce different views and approaches to the world. Combining these views also enriches the scientific toolbox used on a problem, resulting in advances that combine the best of each world and enabling applications that would not be possible otherwise.

Engineers increasingly turn to biologists to understand how living organisms solve problems in the natural environment. Rather than blindly copying nature, such "biomimetic" endeavors often combine the best solutions from nature with artificially engineered components to develop a system that is better for the particular environment than any existing organism. Many significant trends leverage technologies from multiple areas. Smart materials are contributing both function and processing capabilities. Smart systems in turn can be engineered to provide living organisms with bionic capabilities. Living organisms are informing new ways to construct systems (biomimetics) as well as manufacturing components themselves.

Bibliography

Balzani V, Lopez MG, Stoddart J.F. "Molecular Machines". *Acc Chem Res*. 31: 405-14. 1998.

Berg H. C. "Dynamic properties of bacterial flagellar motors". *Nature*. 249: 77-9. 1974.

Drexler Eric. K. "Nanosystems: Molecular Machinery, Manufacturing and Computation: *John Wiley & Sons*. 1992.

Ferguson J.A, Boles TC, Adams CP, Walt DR. "A fiber-optic DNA biosensor microarray for the analysis of gene expression". *Nat Biotechnol*. 1996.

Howard J, Hudspeth AJ, Vale RD. "Movement of microtubules by single kinesin molecules". *Nature*. 1989.

Kitamura K, Tokunaga M, Iwane AH, Yanagida T. "A single myosin head moves along an actin filament with regular steps of 5.3 nanometres". *Nature*. 1999.

Koumura N, Zijlstra RW, van Delden RA, Harada N, Feringa BL. "Light-driven monodirectional molecular rotor". *Nature*. 1999.

Liu H, Schmidt J.J, Bachand G.D, Rizk S.S, Looger L.L, *et al*. "Control of a biomolecular motorpowered nanodevice with an engineered chemical switch". *Nat Mater*. 2002.

Mahadevan L, Matsudaira P. "Motility powered by supramolecular springs and ratchets". *Science*. 2000.

Mao C, Sun W, Shen Z, Seeman N.C. "A nanomechanical device based on the B-Z transition of DNA". *Nature*. 397: 144-6. 1999.

Penrose L.S, "Mechanics of self-reproduction". Ann. Human Genetics 1958.

Pieroni O, Fissi A, Angelini N, Lenci F. "Photoresponsive polypeptides". *Acc Chem Res*. 2001.

Robert A. Freitas Jr. "Nanomedicine". Volume I: Basic Capabilities, Landes Bioscience. Georgetown, TX, 1999.

Seeman N.C. "DNA nanotechnology: novel DNA constructions". *Annu Rev Biophys Biomol Struct*. 1998.

Soong R.K, Bachand G.D, Neves H.P, Olkhovets A.G, Craighead H.G, *et al*. "Powering an inorganic nanodevice with a biomolecular motor". *Science*. 2000.

Yan H, Zhang X, Shen Z, Seeman N.C. "A robust DNA mechanical device controlled by hybridization topology". *Nature*. 2002.

Yasuda R, Noji H, Kinosita K, Jr., Yoshida M. "F1-ATPase is a highly efficient molecular motor that rotates with discrete 120 degree steps". *Cell*. 1998.

Yuqiu J, Juang CB, Keller D, Bustamante C, Beach D, *et al*. "Mechanical, Electrical, and Chemical Manipulation of Single DNA Molecules". *Nanotechnology*. 1992.

Index

Applied Biology 267

Bio-nanotechnology 15
Blow-by-blow account 72

Carbon Nanotube (CNT) 4, 136
Catalyst performance 121
Catalytic nanoparticles 124
Chemical bonds 119
Chemical conversion 120
Cloning 256
Cognitive technology 2
Computational design capabilities 163
Computational Infrastructure 151

Deep-rooted principles 86
Defence Capability 16
Drug Therapies 262

Electron transport phenomena 149
Electron Transport 125
Electron-based tools 163
Electronic Structure Calculations 148
Electronic-structure Methods 150
Energy Conversion 128
Engineered nanomaterials systems 154
Engineering programs 263
Evolutionary change 86

Genetically Modified Organisms (GMOs) 257
Genetics, nanotechnology, and robotics (GNR) 18
Gross domestic product (GDP), 12

High-tech capabilities 17
Human Genome Project (HGP) 7

Incremental nanotechnology 2
Intellectual property 13

Knowledge-base underpinning science 10

Living systems 140

Metalmatrix composite materials 6
Microelectromechanical Systems (MEMS) 144, 271
Monte Carlo simulations 148

Nanomaterials 1
Nanoparticles 65
Nanostructured Materials 139
Nanostructuring semiconductors 129
Nanotechnology Initiative (NNI) 146
Nanotechnology-based manufacturing 5

Organic light emitting diodes (OLEDs) 161
Oxide-Dispersive-Strengthened (ODS) 162

Photolithography 144
Plastic deformation processes 3
Polymerase Chain Reaction (PCR) 272
Professional societies 9

Quantitative measurements 138
Quantum yields (QY) 135

Radical nanotechnology 2
Richard Dawkins 73, 83

Scalable Synthesis 152
Scanning Probe Microscopy (SPM) 123
Silica-based nanomaterial 4
Single-Walled Carbon Nanohorns (SWNHs) 159
Solid-state synthesis approach 157
State-of-the-art experimental research 128

Technological advances 252
Technological possibilities 87
Technological variations 78
Technology revolution 252
Theory, Modelling, and Simulation (TMS) 146
Traditional bulk materials 66
Transmission Electron Microscopy (TEM) 123

Ultrafine-grained structures 3
Unmanned Aerial Vehicles (UAVs) 273

Winning technologies 87